A2-Level

Theide

Editors:
Ellen Bowness, Katie Braid, Joe Brazier, Charlotte Burrows, Katherine Craig, Rosie Gillham, Jane Towle.

Contributors:
Gloria Barnett, Jessica Egan, Derek Harvey, Adrian Schmit, Sophie Watkins.

Proofreaders:
Glenn Rogers, Sue Hocking.

Published by Coordination Group Publications Ltd.

ISBN: 978 1 84762 259 4

With thanks to Laura Stoney for the copyright research.

Data used to construct the graph of daily CO_2 concentration on page 30 reproduced with kind permission from a study at Griffin Forrest, Perthshire performed by the University of Edinburgh and supported by the Natural Environment Research Council.

Data used to construct the graph of yearly CO_2 concentration on page 30 and average yearly CO_2 concentration on page 33 reproduced with kind permission from Atmospheric CO_2 at Mauna Loa Observatory, Scripps Institution of Oceanography, NOAA Earth System Research Laboratory.

Data used to construct the graph of temperature change over the last 1000 years on page 31 reproduced with kind permission from Climate Change 2001: The Scientific Basis, Contribution of Working Group I to the Third Assessment Report of the Intergovernmental Panel on Climate Change, SPM Figure 1. Cambridge University Press.

Data used to construct the graph of methane concentration on page 31 © CSIRO Marine and Atmospheric Research, reproduced with permission from www.csiro.au.

Data used to construct the graph of CO_2 concentration on page 31 reproduced with kind permission from U.S. Global Change Research Program, http://www.usgcrp.gov/usgcrp/nacc/background/scenarios/images/co2hm.gif

Data used to construct the graph of wheat yield on page 32 from Global scale climate-crop yield relationships and the impacts of recent warming. D. B. Lobell and C. B. Field. Environmental Research Letters 2 (2007) 014002 (7pp). IOP Publishing.

Data used to construct the graph of average global temperature on page 33 adapted from Crown Copyright data supplied by the Met Office.

Data used to construct the graph of global sea temperature on page 33 reproduced with kind permission from NASA Goddard Institute for Space Studies.

Diagram showing the distribution of subtropical plankton on page 33 reproduced with kind permission from Plankton distribution changes, due to climate changes – North Sea. (February 2008). In UNEP/GRID-Arendal Maps and Graphics Library http://maps.grida.no/go/graphic/plankton-distribution-changes-due-to-climate-changes-north-sea.

Data used to construct the graph of corn yield on page 33 provided by the U.S. Department of Agriculture – National Agricultural Statistics Service.

Data used to construct the graph showing the stock of spawning cod in the North Sea and the rate of mortality caused by fishing since 1960 on page 41 from the International Council for the Exploration of the Sea.

Data used to construct the graphs on page 105 from S. Hacein–Bey–Abina et al. SCIENCE 302: 415–419 (2003).

Groovy website: www.cgpbooks.co.uk
Jolly bits of clipart from CorelDRAW®
Printed by Elanders Hindson Ltd, Newcastle upon Tyne.

Based on the classic CGP style created by Richard Parsons.

Contents

The Scientific Process

This stuff may look similar to what you learnt at AS, but that's because you need to understand How Science Works for A2 as well. 'How Science Works' is all about the scientific process — how we develop and test scientific ideas. It's what scientists do all day, every day (well, except at coffee time — never come between a scientist and their coffee).

Scientists Come Up with **Theories** — Then **Test Them**...

Science tries to explain **how** and **why** things happen — it **answers questions**. It's all about seeking and gaining **knowledge** about the world around us. Scientists do this by **asking** questions and **suggesting** answers and then **testing** them, to see if they're correct — this is the **scientific process**.

1) **Ask** a question — make an **observation** and ask **why or how** it happens. E.g. why do plants grow faster in glasshouses than outside?

2) **Suggest** an answer, or part of an answer, by forming a **theory** (a possible **explanation** of the observations), e.g. glasshouses are warmer than outside and plants grow faster when it's warmer because the rate of photosynthesis is higher. (Scientists also sometimes form a **model** too — a **simplified picture** of what's physically going on.)

3) Make a **prediction** or **hypothesis** — a **specific testable statement**, based on the theory, about what will happen in a test situation. E.g. the rate of photosynthesis will be faster at 20 °C than at 10 °C.

4) Carry out a **test** — to provide **evidence** that will support the prediction (or help to disprove it). E.g. measure the rate of photosynthesis at various temperatures.

Simone predicted her hair would be worse on date night, based on the theory of sod's law.

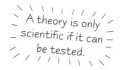

A theory is only scientific if it can be tested.

...Then They **Tell** Everyone About Their **Results**...

The results are **published** — scientists need to let others know about their work. Scientists publish their results in **scientific journals**. These are just like normal magazines, only they contain **scientific reports** (called papers) instead of the latest celebrity gossip.

1) Scientific reports are similar to the **lab write-ups** you do in school. And just as a lab write-up is **reviewed** (marked) by your teacher, reports in scientific journals undergo **peer review** before they're published.

2) The report is sent out to **peers** — other scientists who are experts in the **same area**. They examine the data and results, and if they think that the conclusion is reasonable it's **published**. This makes sure that work published in scientific journals is of a **good standard**.

3) But peer review **can't guarantee** the science is **correct** — other scientists still need to **reproduce** it.

4) Sometimes **mistakes** are made and flawed work is published. Peer review **isn't perfect** but it's probably the best way for scientists to self-regulate their work and to publish **quality reports**.

...Then **Other Scientists** Will **Test** the Theory Too

Other scientists read the published theories and results, and try to **test the theory** themselves. This involves:

• Repeating the **exact same experiments**.
• Using the theory to make **new predictions** and then testing them with **new experiments**.

If the **Evidence** Supports a Theory, It's **Accepted** — for Now

1) If all the experiments in all the world provide good evidence to back it up, the theory is thought of as **scientific 'fact'** (for now).

2) But it will never become **totally indisputable** fact. Scientific **breakthroughs or advances** could provide new ways to question and test the theory, which could lead to **new evidence** that **conflicts** with the current evidence. Then the testing starts all over again...

And this, my friend, is the **tentative nature of scientific knowledge** — it's always **changing** and **evolving**.

The Scientific Process

So scientists need evidence to back up their theories. They get it by carrying out experiments, and when that's not possible they carry out studies. But why bother with science at all? We want to know as much as possible so we can use it to try and improve our lives (and because we're nosy).

Evidence Comes from Lab Experiments...

1) Results from **controlled experiments** in **laboratories** are great.

2) A lab is the easiest place to **control variables** so that they're all **kept constant** (except for the one you're investigating).

3) This means you can draw meaningful **conclusions**.

> For example, if you're investigating how light intensity affects the rate of photosynthesis you need to keep everything but the light intensity constant, e.g. the temperature, the concentration of carbon dioxide etc.

...and Well-Designed Studies

1) There are things you **can't** investigate in a lab, e.g. whether using a pesticide on farmland affects the number of non-pest species. You have to do a study instead.

2) You still need to try and make the study as controlled as possible to make it **more reliable**. But in reality it's **very hard** to control **all the variables** that **might** be having an effect.

3) You can do things to help, like having a **control** — e.g. an area of similar farmland nearby where the pesticide isn't applied. But you can't easily rule out every possibility.

Having a control reduced the effect of exercise on the study.

See pages 106-108 for more on study design.

Society Makes Decisions Based on Scientific Evidence

1) Lots of scientific work eventually leads to **important discoveries** or breakthroughs that could **benefit humankind**.

2) These results are **used by society** (that's you, me and everyone else) to **make decisions** — about the way we live, what we eat, what we drive, etc.

3) All sections of society use scientific evidence to make decisions, e.g. politicians use it to devise policies and individuals use science to make decisions about their own lives.

Other factors can **influence** decisions about science or the way science is used:

Economic factors

- Society has to consider the **cost** of implementing changes based on scientific conclusions — e.g. the NHS can't afford the most expensive drugs without **sacrificing** something else.

- Scientific research is **expensive** so companies won't always develop new ideas — e.g. developing new drugs is costly, so pharmaceutical companies often only invest in drugs that are likely to make them **money**.

Social factors

- **Decisions** affect **people's lives** — E.g. scientists may suggest **banning smoking** and **alcohol** to prevent health problems, but shouldn't **we** be able to **choose** whether **we** want to smoke and drink or not?

Environmental factors

- Scientists believe **unexplored regions** like remote parts of rainforests might contain **untapped drug** resources. But some people think we shouldn't **exploit** these regions because any interesting finds may lead to **deforestation** and **reduced biodiversity** in these areas.

So there you have it — how science works...

Hopefully these pages have given you a nice intro to how science works, e.g. what scientists do to provide you with 'facts'. You need to understand this, as you're expected to know how science works — for the exam and for life.

Populations and Ecosystems

Now, I need something momentous to say at the start of this book — these pages are one small step for an A2 level biology student, but one giant leap for the subject of biology... oh, someone's said that before. Nuts.

You Need to **Learn Some Definitions** to get you **Started**

Habitat	—	The **place** where an organism **lives**, e.g. a rocky shore or a field.
Population	—	**All** the organisms of **one species** in a **habitat**.
Community	—	Populations of **different species** in a habitat make up a **community**.
Ecosystem	—	**All** the **organisms** living in a **particular area** and all the **non-living** (abiotic) conditions, e.g. a freshwater ecosystem such as a lake.
Abiotic conditions	—	The **non-living** features of the ecosystem, e.g. **temperature** and **availability of water**.
Biotic conditions	—	The **living** features of the ecosystem, e.g. the presence of **predators** or **food**.
Niche	—	The **role** of a species within its habitat, e.g. what it eats, where and when it feeds.
Adaptation	—	A **feature** that members of a species have that **increases** their chance of **survival** and **reproduction**, e.g. **giraffes** have **long necks** to help them reach vegetation that's high up. This increases their chances of survival when food is **scarce**.

Being a member of the undead made it hard for Mumra to know whether he was a living or a non-living feature of the ecosystem.

Every Species Occupies a Different Niche

1) The **niche** a species occupies within its habitat includes:

   - Its **biotic** interactions — e.g. the organisms it **eats**, and those it's **eaten by**.
   - Its **abiotic** interactions — e.g. the **oxygen** an organism breathes in, and the **carbon dioxide** it breathes out.

   Don't get confused between habitat (where a species lives) and niche (what it does in its habitat).

2) Every species has its own **unique niche** — a niche can only be occupied by **one species**.

3) It may **look** like **two species** are filling the **same niche** (e.g. they're both eaten by the same species), but there'll be **slight differences** (e.g. variations in what they eat).

4) If two species **try** to occupy the **same niche**, they will **compete** with each other. One species will be **more successful** than the other, until **only one** of the species is **left**.

5) Here are a couple of examples of niches:

   ### Common pipistrelle bat
   This bat lives throughout Britain on **farmland**, **open woodland**, **hedgerows** and **urban areas**. It feeds by **flying** and catching **insects** using echolocation (**high-pitched sounds**) at a **frequency** of around **45 kHz**.

   ### Soprano pipistrelle bat
   This bat lives in Britain in **woodland** areas, close to **lakes** or **rivers**. It feeds by **flying** and catching **insects** using **echolocation**, at a **frequency** of **55 kHz**.

   It may **look like** both species are filling the **same niche** (e.g. they both eat insects), but there are **slight differences** (e.g. they use **different frequencies** for their echolocation).

Populations and Ecosystems

Organisms are Adapted to Biotic and Abiotic Conditions

1) As you know, **adaptations** are features that **increase** an organism's chance of **survival** and **reproduction**.

2) They can be **physiological** (processes **inside** their body), **behavioural** (the way an organism **acts**) or **anatomical** (**structural features** of their body).

3) Organisms with better adaptations are **more likely** to **survive**, **reproduce** and **pass on** the alleles for their adaptations, so the adaptations become **more common** in the population. This is called **natural selection**.

4) Every species is adapted to **use** an **ecosystem** in a way that **no other** species can. For example, only giant anteaters can **break into** ant nests and **reach** the ants. They have **claws** to rip open the nest, and a **long, sticky tongue** which can move **rapidly** in and out of its mouth to **pick up** the ants.

5) Organisms are **adapted** to both the **abiotic conditions** (e.g. how much **water** is available) and the **biotic conditions** (e.g. what **predators** there are) in their ecosystem.

Here are a few ways that **different organisms** are **adapted** to the **abiotic** or the **biotic** conditions in their ecosystems:

Adaptations to abiotic conditions

- **Otters** have **webbed paws** — this means they can both **walk** on land and **swim** effectively. This increases their chance of survival because they can **live** and **hunt** both on land and in water.

- **Whales** have a **thick layer** of **blubber** (fat) — this helps to keep them **warm** in the **coldest seas**. This increases their chance of survival because they can **live** in places where food is plentiful.

- **Brown bears hibernate** — they **lower their metabolism** (all the chemical reactions taking place in their body) over **winter**. This increases their chance of survival because they can **conserve energy** during the **coldest** months.

Adaptations to biotic conditions

- **Sea otters** use **rocks** to **smash open** shellfish and clams. This increases their chance of survival because it gives them **access** to **another source** of food.

- **Scorpions dance** before **mating** — this makes sure they **attract a mate** of the **same species**. This increases their chance of reproduction by making **successful mating** more likely.

- Some **bacteria** produce **antibiotics** — these **kill other species** of bacteria in the **same area**. This increases their chance of survival because there's **less competition** for **resources**.

Take your partner 1, 2, 3, swing them round a sycamore tree.

Practice Questions

Q1 What is the name given to all the organisms of one species in a habitat?

Q2 Define a community.

Q3 Give the term for the non-living features of an ecosystem.

Q4 What happens when two species try to occupy the same niche in an ecosystem?

Exam Question

Q1 Common pipistrelle bats have light, flexible wings, which means they can fly fast and are manoeuvrable. They hunt insects at night using echolocation and live on farmland, in open woodland, hedgerows and urban areas. They make unique mating calls to find mates, hibernate through the winter, and roost in cracks in trees and buildings during the day.

   a) Describe the habitat of the common pipistrelle bat. [2 marks]

   b) Explain how the common pipistrelle bat is adapted to the biotic conditions in its ecosystem. [3 marks]

Unique quiche niche — say it ten times really fast...

All this population and ecosystem stuff is pretty wordy I'm afraid, but I'll tell you what, you'll be missing it when you get onto the really sciencey stuff later. You just need to learn and re learn all the key words here, then when they ask you to interpret some bat-related babble in the exam, you'll know exactly what they're talking about. Niche work.

Investigating Populations

Examiners aren't happy unless you're freezing to death in the rain in a field somewhere in the middle of nowhere. Still, it's better than being stuck in the classroom being bored to death learning about fieldwork techniques...

You need to be able to **Investigate Populations** of **Organisms**

Investigating **populations** of organisms involves looking at the **abundance** and **distribution** of **species** in a particular **area**.

1) **Abundance** — the **number of individuals** of **one species** in a **particular area**.
   The abundance of **mobile organisms** and **plants** can be estimated by simply counting the **number** of individuals in samples taken. There are other measures of abundance that can be used too:
   - **Frequency** — the **number of samples** a species is **recorded in**, e.g. 70% of samples.
   - **Percentage cover** (for plants only) — **how much** of the area you're investigating is **covered** by a species.

2) **Distribution** — this is **where** a particular species is within the **area you're investigating**.

You need to take a **Random Sample** from the **Area You're Investigating**

Most of the time it would be too **time-consuming** to measure the **number of individuals** and the **distribution** of every species in the **entire area** you're investigating, so instead you take **samples**:

1) **Choose** an **area** to sample — a **small** area **within** the area being investigated.

2) Samples should be **random** to **avoid bias**, e.g. if you were investigating a field you could pick random sample sites by dividing the field into a **grid** and using a **random number generator** to select **coordinates**.

3) Use an **appropriate technique** to take a sample of the population (see below and the next page).

4) **Repeat** the process, taking as many samples as possible. This gives a more **reliable** estimate for the **whole area**.

5) The **number of individuals** for the **whole area** can then be **estimated** by taking an **average** of the data collected in each sample and **multiplying** it by the size of the whole area. The **percentage cover** for the whole area can be estimated by taking the average of all the samples.

Different Methods are Used to **Investigate Different Organisms**

① **Pitfall Traps** and **Pooters** are used to **Investigate Ground Insects**

Pitfall traps

1) **Pitfall traps** are **steep-sided containers** that are sunk in a **hole** in the ground. The top is **partially open**.

2) Insects **fall** into the container and **can't get out** again — they're **protected** from **rain** and **some predators** by a **raised lid**.

3) The sample can be affected by **predators small enough** to fall into the pitfall trap though — they may **eat other insects**, **affecting the results**.

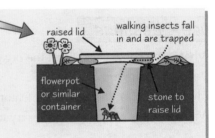

raised lid · walking insects fall in and are trapped · flowerpot or similar container · stone to raise lid

Pooters

1) **Pooters** are **jars** that have **rubber bungs** sealing the top, and **two tubes** stuck through the bung.

2) The **shorter tube** has **mesh** over the end that's in the jar. The **longer tube** is **open** at both ends.

3) When you **inhale** through the shorter tube, **air is drawn** through the longer tube. If you **place** the end of the **longer tube** over an insect it'll be **sucked** into the jar.

4) It can take a **long time** (or **lots of people**) to get a **large sample** using pooters. Some species may be **missed** if the sample **isn't large enough**.

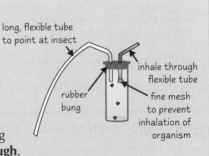

long, flexible tube to point at insect · inhale through flexible tube · rubber bung · fine mesh to prevent inhalation of organism

Investigating Populations

② Quadrats and Transects are used to Investigate Plant Populations

Quadrats

1) A **quadrat** is a **square** frame divided into a **grid** of
   100 **smaller squares** by strings attached across the frame.

2) Quadrats are **placed on the ground** at **different points**
   within the area you're investigating.

3) The **species frequency** or the **number of individuals**
   of each species is recorded in **each quadrat**.

4) The **percentage cover** of a species can also be measured
   by counting how much of the quadrat is **covered** by the
   species — you count a square if it's **more than half-covered**.
   Percentage cover is a **quick** way to investigate populations
   and you **don't** have to **count** all the **individual** plants.

5) Quadrats are useful for **quickly** investigating areas with
   plant species that **fit** within a **small quadrat** — areas with
   larger plants and **trees** need **very large** quadrats.

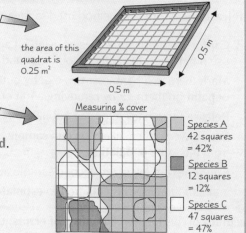

the area of this quadrat is 0.25 m²

0.5 m

0.5 m

Measuring % cover

Species A
42 squares
= 42%

Species B
12 squares
= 12%

Species C
47 squares
= 47%

Transects

You can use **lines** called **transects** to help find out how plants are **distributed across** an area,
e.g. how species change from a hedge towards the middle of a field. There are three types:

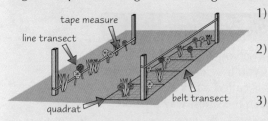

tape measure

line transect

quadrat

belt transect

1) **Line transects** — a **tape measure** is placed **along** the transect
   and the species that **touch** the tape measure are **recorded**.

2) **Belt transects** — quadrats are placed next to each other **along**
   the transect to work out **species frequency** and **percentage cover**
   along the transect.

3) **Interrupted transects** — instead of investigating the whole transect
   of either a line or a belt, you can take **measurements** at **intervals**.

③ Beating Trays are used to Investigate Insects Found in Vegetation

Beating trays are usually white so you can see the insects.

1) A **beating tray** is a **tray** or **sheet** held **under** a plant or tree.

2) The plant or tree is **shaken** and a sample of insects **falls onto** the beating tray.

3) You can take **large samples** using beating trays, giving **good estimates** of the **abundance** of each species.

4) However, the **sample** may **not be random** because most of it will be made up of insects that **fall easily**
   when the vegetation is **shaken**.

Practice Questions

Q1 Explain why samples of a population are taken.

Q2 Give one drawback of using beating trays to investigate insect populations.

Exam Question

Q1 A student wants to sample a population of daffodils in a field.

   a) How could she avoid bias in her investigation? [1 mark]

   b) Describe how she could investigate the percentage cover of daffodils in the field. [4 marks]

Beating trays — as used by wrestlers on TV...

*There are plenty of pitfalls to avoid when investigating populations, but if you give these pages a good read you'll leap
straight over them. Just be aware that every technique for collecting data on populations has its drawbacks.*

Investigating Populations

More practical fun... no wait, there's some data interpretation on these pages too. Sadly, dealing with data is pretty important — I'd bet my phone, car and beloved bike that there'll be at least one data interpretation question in the exam

Mark-Release-Recapture is Used to Investigate More Mobile Species

Mark-release-recapture is a method used to measure the **abundance** of more **mobile** species. Here's how it's done:

1) **Capture** a sample of a species using an **appropriate technique**, e.g. you could use pitfall traps to capture mobile ground insects (see p. 6), and **count** them.

2) **Mark** them in a harmless way, e.g. by putting a spot of **paint** on them, or by **removing** a tuft of **fur**.

3) **Release** them back into their habitat.

4) Wait a week, then take a **second sample** from the **same population**.

5) **Count** how many of the second sample **are marked**.

6) You can then use this **equation** to **estimate** the **total** population size.

$$\text{Total population size} = \frac{\text{Number caught in 1st sample} \times \text{Number caught in 2nd sample}}{\text{Number marked in 2nd sample}}$$

The **accuracy** of this method (how **free of errors** it is) depends on a few **assumptions**:

1) The marked sample has had enough **time** and **opportunity** to **mix** back in with the population.

2) The marking hasn't affected the individuals' **chances of survival**, and is **still visible**.

3) **Changes** in **population size** due to **births**, **deaths** and **migration** are **small** during the period of the study.

You Need to Carry Out a Risk Assessment for all Practical Work

When you're carrying out fieldwork to investigate populations you expose yourself to **risks** — things that could **potentially** cause you **harm**. You need to think about **what risks** you'll be exposed to during fieldwork, so you can **plan** ways to **reduce** the **chance** of them happening — this is called a **risk assessment**. Risk assessments are always carried out to ensure that fieldwork's done in the **safest way possible**.

Here are some **examples** of the fieldwork risks when investigating populations and the **ways** to reduce the risks:

Falls and slips	Wear suitable footwear for the terrain, e.g. wellies on wet or boggy ground, and take care on rough terrain. Make sure the study area isn't near any cliffs or on steep ground.
Bad weather	Check the weather forecast beforehand and take precautions, e.g. wear warm or waterproof clothing on cold or wet days. If the weather is too bad, do the fieldwork another day.
Stings and bites	Wear insect repellent or, if you have an allergy, take medication with you.

OK chaps, just get out your lightning-proof suits and let's crack on with the fieldwork.

There are Ethical Issues to Consider When Doing Fieldwork

All fieldwork **affects** the **environment** where it's carried out, e.g. lots of people **walking around** may cause **soil erosion**. Some people don't think it's **right** to **damage** the **environment** when doing fieldwork, so investigations should be planned to have the **smallest impact possible**, e.g. people should restrict **where they walk** to the area being studied.

Some fieldwork **affects** the **organisms** being studied, e.g. **capturing** an organism for study may cause it **stress**. Some people don't think it's **right** to **distress** organisms **at all** when doing fieldwork, so investigations should be planned so that organisms are treated with **great care**, and are **kept** and **handled** as **little** as possible. They should also be **released** as soon as possible after they have been captured.

Investigating Populations

You need to *Analyse* and *Interpret Data* on the *Distribution of Organisms*

Here's an **example** of the kind of thing you might get in the **exam**:

A group of students investigated how the **distribution** of plant species changed with **distance** from a path. They used a **belt transect** (see p. 7) and measured **percentage cover** of plant species in each quadrat. The students also carried out a **survey** at the **same** location to record how many people **strayed away** from the path, and **how far** they strayed.

Here's a **table** and a **graph** showing their results.

You might have to:

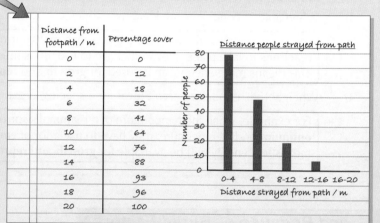

Distance from footpath / m	Percentage cover
0	0
2	12
4	18
6	32
8	41
10	64
12	76
14	88
16	93
18	96
20	100

1) **Describe the data:**

- The table shows **low** percentage cover of plants **near** the path, e.g. **2 m** from the path it was **12%**, but **higher** percentage cover **away** from the path, e.g. **20 m** from the path it was **100%**.

- The graph shows **lots of walkers near** the path, e.g. **0-4 m** from the path there were **79** walkers, but **fewer** walkers away from the path, e.g. **16-20 m** from the path there **were none**.

2) **Draw conclusions:**

- There's a **positive correlation** between **distance** from the path and **percentage cover** of plants — as distance from the path **increases**, the percentage cover of plants **increases**.

- There's a **negative correlation** between **distance** from the path and the **number** of people that walk there — as distance from the path **increases**, the number of people that walk there **decreases**.

- There's a **negative correlation** between the **number of walkers** and the **percentage cover** of plants — the **higher** the number of people that walk over an area, the **lower** the percentage cover of plants.

You **can't conclude** that the **lower percentage cover** of plants **near** the path is **caused** by the **higher number** of **people** walking there. There could be **other factors** involved that affect the percentage cover of plants, e.g. the path may be covered by **stones** or **gravel**, so plants won't grow **on** or **near** the path regardless of how many people walk on it.

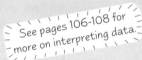
See pages 106-108 for more on interpreting data.

3) **Suggest explanations for your conclusions:**

As you move **away** from the path the **number** of people that trample the ground **decreases** because people tend to **follow the path**. As you move **away** from the path the **percentage cover** of plants **increases** because plants **grow** and **survive better** where they're trodden on **less**.

Practice Questions

Q1   Name one fieldwork risk when investigating populations.

Q2   Give an example of an ethical issue associated with fieldwork into populations.

Exam Question

Number of snails caught in first sample	Number of snails caught in second sample	Number of marked snails caught in second sample
52	38	14

Q1   The size of a snail population was investigated using the mark-release-recapture method. The table shows the results.

a) Describe the method that could have been used to collect the data.   [5 marks]

b) Calculate the total population size.   [2 marks]

Risks associated with this book — laughter, increased intelligence...

Mark-release-recapture isn't too bad — it's exactly what it sounds like. Risk assessments aren't too bad either, they usually just involve a bit of common-sense thinking to work out what might be dangerous. As always, interpreting data can be a pain — just make sure you're clear that even if two things correlate, it doesn't mean that one is caused by the other.

Variation in Population Size

Uh-oh, anyone who loves cute little bunny-wunnys look away now — these pages are about how the population sizes of organisms fluctuate and the reasons why. One of the reasons, I'm sad to say, is because the little rabbits get eaten.

Population Size Varies Because of Abiotic Factors...

Remember — abiotic factors are the non-living features of the ecosystem.

1) **Population size** is the **total number** of organisms of **one species** in a **habitat**.

2) The **population size** of any species **varies** because of **abiotic** factors, e.g. the amount of **light**, **water** or **space** available, the **temperature** of their surroundings or the **chemical composition** of their surroundings.

3) When abiotic conditions are **ideal** for a species, organisms can **grow fast** and **reproduce successfully**.

> For example, when the temperature of a mammal's surroundings is the ideal temperature for **metabolic reactions** to take place, they don't have to **use up** as much energy **maintaining** their **body temperature**. This means more energy can be used for **growth** and **reproduction**, so their population size will **increase**.

4) When abiotic conditions **aren't ideal** for a species, organisms **can't** grow as **fast** or reproduce as **successfully**.

> For example, when the temperature of a mammal's surroundings is significantly **lower** or **higher** than their **optimum** body temperature, they have to **use** a lot of **energy** to maintain the right **body temperature**. This means less energy will be available for **growth** and **reproduction**, so their population size will **decrease**.

...and Because of Biotic Factors

Remember — biotic factors are the living features of the ecosystem.

(1) Interspecific Competition — Competition Between Different Species

1) Interspecific competition is when organisms of **different species compete** with each other for the **same resources**, e.g. **red** and **grey** squirrels compete for the same **food sources** and **habitats** in the **UK**.

2) Interspecific competition between two species can mean that the **resources available** to **both** populations are **reduced**, e.g. if they share the **same** source of food, there will be **less** available to both of them. This means both populations will be **limited** by a lower amount of food. They'll have less **energy** for **growth** and **reproduction**, so the population sizes will be **lower** for both species. E.g. in areas where both **red** and **grey** squirrels live, both populations are **smaller** than they would be if there was **only one** species there.

3) If **two** species are competing but one is **better adapted** to its surroundings than the other, the less well adapted species is likely to be **out-competed** — it **won't** be able to **exist** alongside the better adapted species. E.g. since the introduction of the **grey squirrel** to the UK, the native **red squirrel** has **disappeared** from large areas. The grey squirrel has a better chance of **survival** because it's **larger** and can store **more fat** over winter. It can also eat a **wider range** of **food** than the red squirrel.

Never mind what the doctors said, Nutkin knew his weight problem would increase his chance of survival.

(2) Intraspecific Competition — Competition Within a Species

Intraspecific competition is when organisms of the **same species compete** with each other for the **same resources**.

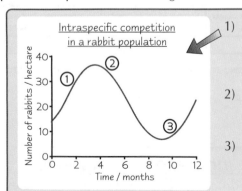

1) The **population** of a species **increases** when resources are **plentiful**. As the population increases, there'll be **more** organisms competing for the **same amount** of **space** and **food**.

2) Eventually, resources such as food and space become **limiting** — there **isn't enough** for all the organisms. The population then begins to **decline**.

3) A **smaller** population then means that there's **less competition** for space and food, which is **better** for **growth** and **reproduction** — so the population starts to **grow** again.

Variation in Population Size

③ Predation — Predator and Prey Population Sizes are Linked

Predation is where an organism (the predator) kills and eats another organism (the prey),
e.g. lions kill and eat (**predate** on) buffalo. The **population sizes** of predators and prey are
interlinked — as the population of one **changes**, it **causes** the other population to **change**:

1) As the **prey** population **increases**, there's **more food** for predators, so the **predator** population **grows**. E.g. in the graph on the right the **lynx** population **grows** after the **snowshoe hare** population has **increased** because there's **more food** available.

2) As the **predator** population **increases**, **more prey** is **eaten** so the **prey** population then begins to **fall**. E.g. **greater numbers** of lynx eat lots of snowshoe hares, so their population **falls**.

3) This means there's **less food** for the **predators**, so their population **decreases**, and so on. E.g. **reduced** snowshoe hare numbers means there's **less food** for the lynx, so their population **falls**.

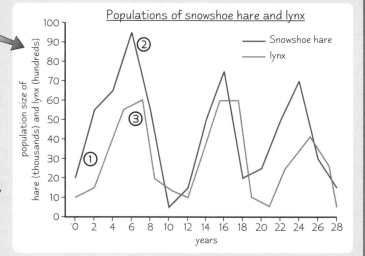

Predator-prey relationships are usually more **complicated** than this though because there are **other factors** involved, like availability of **food** for the **prey**. E.g. it's thought that the population of snowshoe hare initially begins to **decline** because there's **too many** of them for the amount of **food available**. This is then **accelerated** by **predation** from the lynx.

Practice Questions

Q1 Give one example of how an abiotic factor can affect population size.

Q2 What is interspecific competition?

Q3 What will be the effect of interspecific competition on the population size of a species?

Q4 What does it mean when a species is out-competed?

Q5 Give one example of interspecific competition.

Q6 Define intraspecific competition.

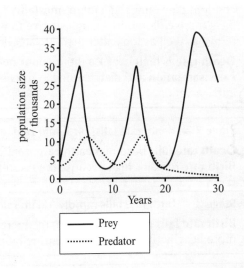

Exam Question

Q1 The graph on the right shows the population size of a predator species and a prey species over a period of 30 years.

a) Using the graph, describe and explain how the population sizes of the predator and prey species vary over the first 20 years. [7 marks]

b) The numbers of species B declined after year 20 because of a disease. Describe and explain what happened to the population of species A. [4 marks]

Predator-prey relationships — they don't usually last very long...

You'd think they could have come up with names a little more different than inter- and intraspecific competition. I always remember it as int-er means diff-er-ent species. The factors that affect population size are divided up nicely for you here — abiotic factors, competition and predation — just like predators like to nicely divide up their prey into bitesize chunks.

Human Populations

These pages are about how human populations change, so they're about joyful births... and not so joyful deaths.

Human Population Growth is Calculated using Birth and Death Rates

Human population sizes constantly **change**. Whether they're **growing** or **shrinking** (and by **how much**) depends on the population's **birth rate** and **death rate**.

1) **Birth rate** — the number of **live births each year** for **every 1000** people in the population, e.g. a birth rate of **10/1000** would mean that in one year there were **10 live births** for every **1000 people**.

2) **Death rate** — the number of people that **die each year** for every **1000** people in the population, e.g. a death rate of **10/1000** would mean that in one year there were **10 deaths** for every **1000 people**.

You can work out **how fast** the population's **changing** by calculating the **population growth rate**:

Population growth rate is how much the **population** size **increases** or **decreases** in a **year**. You can work it out using the **birth** and **death rate**:

$$\text{population growth rate (per 1000 people per year)} = \text{birth rate} - \text{death rate}$$

This gives you the **overall (net) number of people** that the population **grows** or **shrinks by** in a **year** for every **1000 people**. For example, if the birth rate was **13/1000** and the death rate was **10/1000** the population would grow by **3 people** for every **1000 people each year** (or **3/1000** people per year). It's normally given as a **percentage**, so a growth rate of **3/1000** people per year would be **0.3**% (3/1000 × 100%).

$$3/1000 = 13/1000 - 10/1000$$
$$3/1000 \times 100\% = 0.3\%$$

The Demographic Transition Model shows Trends in Human Populations

The **Demographic Transition Model** (**DTM**) is a graph that shows changes in **birth rate**, **death rate** and **total population size** for a **human population** over a **long period** of time. It's divided into **five** stages:

<u>Stage 1</u> — birth rate and death rate fluctuate at a **high level**. The population stays **low**.

Birth rate is high because there's **no birth control** or **family planning** and **education** is poor. Lots of children **die young** (high **infant mortality**), so parents have more children so enough **survive** to **work** on farms, as well as **look after** them in later life.

Death rate is high because there's **poor health care**, **sanitation** and **diet**, leading to disease and starvation.

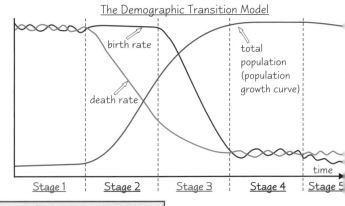

The Demographic Transition Model

birth rate

death rate

total population (population growth curve)

time

Stage 1 | Stage 2 | Stage 3 | Stage 4 | Stage 5

<u>Stage 2</u> — death rate **falls**, birth rate **remains high**. The population **increases rapidly**.
Death rate falls because **health care**, **sanitation** and **diet improve**.
Birth rate remains **high** because there's still little **birth control** or **family planning**.

'Demographic' means it's to do with human populations.

<u>Stage 3</u> — birth rate falls rapidly, death rate falls more slowly. The population increases at a slower rate.
Birth rate falls rapidly because of the increased use of birth control and family planning. Also, the economy becomes more heavily based on **manufacturing** rather than agriculture, so fewer children are needed to work on farms.

<u>Stage 4</u> — birth rate and death rate fluctuate at a **low level**. The population remains **stable** but **high**.
Birth rate **stays low** because there's an **increased demand** for **luxuries** and **material possessions**, so **less money** is **available** to raise children. They're not needed to work to **provide income**, so parents have **fewer children**.

<u>Stage 5</u> — birth rate begins to **fall**, death rate **remains stable**. The population begins to **decrease**.
Birth rate falls because children are **expensive** to raise and people often have **dependent elderly relatives**.
Death rate remains **steady** despite continued health care advances as **larger generations** of elderly people die.

Human Populations

Human Population Data can be Plotted in Different Ways

① Population Growth Curves show Change in Population Size

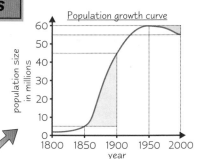

Population change can be shown by a **population growth curve** (the **DTM** has one, see previous page). They're made by plotting data for **population size** against **time**.

1) **Growth curves** show whether the population was **increasing** or **decreasing** by the direction of the curve (**up or down**).

2) The **steepness** of the curve shows **how fast** the population was **changing** (the **steeper** the curve, the **faster** it was changing). You can use the curve to calculate the **rate of change**. For example, between **1850** and **1900** this population **increased** from 5 to **45** million. An increase of **40** million in **50** years meant the population **increased** at a rate of **800 000 people per year** (40 000 000 ÷ 50 = 800 000). Between **1950** and **2000**, the population **decreased** by **5** million in **50** years, so it **decreased** at a rate of **100 000 people per year** (5 000 000 ÷ 50 = 100 000).

② Survival Curves show Survival Rates

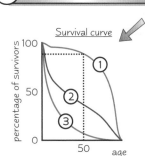

Survival curves show the **percentage** of all the individuals that were **born** in a population that are still **alive** at any **given age**. This gives a **survival rate** for any given age, e.g. population 1 has a survival rate of around **90%** for individuals at the age of **50** — 90% of people survive to be 50.

> Population 1 — **few** people die at a young age, **lots** of people **survive** to an old age.
>
> Population 2 — **many** people die at a young age, but **some survive** to an old age.
>
> Population 3 — **most people die** at a young age, very **few survive** to an old age.

Life expectancy is the **age** that a person born into a population is expected to **live to** — it's worked out by calculating the **average age** that people **die**.

③ Age-sex Pyramids show Population Structure

Population structure can be shown using **age-sex pyramids**. These show how many **males** and **females** there are in different **age groups** within a population.

1) This population has **a lot of young people** with **very few** surviving to **old age** — there's a **low** life expectancy (**DTM stage 1**).

2) This population has **a lot** of **young people** with **more surviving** to **old age** — life expectancy is **higher** (**DTM stage 2**).

3) This population has **fewer young people** with **a lot** of **older people** — life expectancy is **high** (**DTM stage 5**).

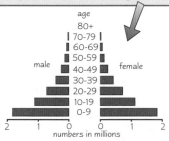

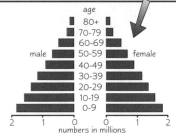

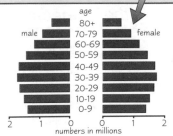

Practice Questions

Q1 How do you calculate population growth rate from birth and death rate?

Q2 What is shown by an age-sex pyramid?

Exam Question

Q1 Describe the differences in population size and structure at stage 1 compared to stage 5 of the DTM. [4 marks]

Population's growth rate — almost 20 cm a year now he's a teenager...

Boy, when it comes to human populations these biologists love their graphs. Even if you feel like your brain's turning to custard, you need to understand what the graphs are showing — you might have to interpret them in the exam.

Photosynthesis, Respiration and ATP

All organisms need energy for life processes (and you'll need some for revising A2 Biology), so it's pretty important stuff. Annoyingly, it's pretty complicated stuff too, but 'cos I'm feeling nice today we'll take it slowly, one bit at a time...

Biological Processes Need Energy

Plant and animal cells **need energy** for biological processes to occur:

- **Plants** need energy for things like **photosynthesis**, **active transport** (e.g. to take in minerals via their roots), **DNA replication**, **cell division** and **protein synthesis**.
- **Animals** need energy for things like **muscle contraction**, maintenance of **body temperature**, **active transport**, **DNA replication**, **cell division** and **protein synthesis**.

Without energy, these biological processes would stop and the plant or animal would die.

Photosynthesis Stores Energy in Glucose

1) **Photosynthesis** is the process where **energy** from **light** is used to **make glucose** from H_2O and CO_2 (the light energy is **converted** to **chemical energy** in the form of glucose).

2) Photosynthesis occurs in a **series** of **reactions**, but the overall equation is:

$$6CO_2 + 6H_2O + \text{Energy} \Longrightarrow C_6H_{12}O_6 \text{ (glucose)} + 6O_2$$

3) So, energy is **stored** in the **glucose** until the plants **release** it by **respiration**.

4) Animals obtain glucose by **eating plants** (or **other animals**), then respire the glucose to release energy.

Cells Release Energy from Glucose by Respiration

1) **Plant** and **animal** cells **release energy** from **glucose** — this process is called **respiration**.

2) This energy is used to power all the **biological processes** in a cell.

3) There are two types of respiration:
- **Aerobic respiration** — respiration **using oxygen**.
- **Anaerobic respiration** — respiration **without oxygen**.

4) Aerobic respiration produces **carbon dioxide** and **water**, and releases **energy**. The overall equation is:

$$C_6H_{12}O_6 \text{ (glucose)} + 6O_2 \Longrightarrow 6CO_2 + 6H_2O + \text{Energy}$$

ATP is the Immediate Source of Energy in a Cell

1) A cell **can't** get its energy **directly** from glucose.

2) So, in respiration, the **energy released** from glucose is used to **make ATP** (adenosine triphosphate). ATP is made from the nucleotide base **adenine**, combined with a **ribose sugar** and **three phosphate groups**.

3) It **carries energy** around the cell to where it's **needed**.

4) **ATP** is **synthesised** from **ADP** and **inorganic phosphate** (P_i) using energy from an **energy-releasing** reaction, e.g. the **breakdown** of **glucose** in **respiration**. The energy is stored as **chemical energy** in the **phosphate bond**. The enzyme **ATP synthase** catalyses this reaction.

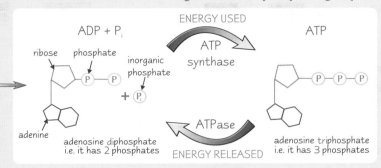

5) ATP **diffuses** to the part of the cell that **needs** energy.

6) Here, it's **broken down** back into **ADP** and **inorganic phosphate** (P_i). Chemical **energy** is **released** from the phosphate bond and used by the cell. **ATPase** catalyses this reaction.

7) The ADP and inorganic phosphate are **recycled** and the process starts again.

Inorganic phosphate (P_i) is just the fancy name for a single phosphate.

Photosynthesis, Respiration and ATP

ATP has Specific Properties that Make it a Good Energy Source

1) ATP stores or releases only a **small**, **managable amount** of energy at a time, so **no** energy is **wasted**.
2) It's a **small**, **soluble** molecule so it can be **easily transported** around the cell.
3) It's **easily broken down**, so energy can be **easily released**.
4) It can **transfer energy** to another molecule by transferring one of its **phosphate groups**.
5) ATP **can't pass out** of the **cell**, so the cell **always** has an immediate supply of energy.

Karen needed a lot
of energy just to keep
her headdress on...

You Need to Know Some Basics Before You Start

There are some pretty confusing technical terms in this section that you need to get your head around:

- **Metabolic pathway** — a **series** of **small reactions** controlled by **enzymes**, e.g. **respiration** and **photosynthesis**.
- **Phosphorylation** — **adding phosphate** to a molecule, e.g. **ADP** is phosphorylated to **ATP** (see previous page).
- **Photophosphorylation** — **adding phosphate** to a molecule using **light**.
- **Photolysis** — the **splitting** (lysis) of a molecule using **light** (photo) energy.
- **Hydrolysis** — the **splitting** (lysis) of a molecule using **water** (hydro).
- **Decarboxylation** — the **removal** of **carbon dioxide** from a molecule.
- **Dehydrogenation** — the **removal** of **hydrogen** from a molecule.
- **Redox reactions** — reactions that involve **oxidation** and **reduction**.

> **Remember redox reactions:**
> 1) If something is **reduced** it has **gained electrons** (e^-), and may have **gained hydrogen** or lost oxygen.
> 2) If something is **oxidised** it has **lost electrons**, and may have **lost hydrogen** or gained oxygen.
> 3) Oxidation of one molecule **always** involves reduction of another molecule.

One way to remember electron and hydrogen movement is OILRIG. Oxidation Is Loss, Reduction Is Gain.

Photosynthesis and Respiration Involve Coenzymes

1) A **coenzyme** is a molecule that **aids** the **function** of an **enzyme**.
2) They work by **transferring** a **chemical group** from one molecule to another.
3) A coenzyme used in **photosynthesis** is **NADP**. NADP transfers **hydrogen** from one molecule to another — this means it can **reduce** (give hydrogen to) or **oxidise** (take hydrogen from) a molecule.
4) Examples of coenzymes used in **respiration** are: **NAD**, **coenzyme A** and **FAD**.
   - NAD and FAD transfer **hydrogen** from one molecule to another — this means they can **reduce** (give hydrogen to) or **oxidise** (take hydrogen from) a molecule.
   - **Coenzyme A** transfers **acetate** between molecules (see pages 23-24).

When hydrogen is transferred between molecules, electrons are transferred too.

Practice Questions

Q1 Write down three biological processes in animals that need energy.
Q2 What is photosynthesis?
Q3 What is the overall equation for aerobic respiration?
Q4 How many phosphate groups does ATP have?
Q5 Give the name of a coenzyme involved in photosynthesis.

Exam Question

Q1 ATP is the immediate source of energy inside a cell.
Describe how the synthesis and breakdown of ATP meets the energy needs of a cell. [6 marks]

Oh dear, I've used up all my ATP on these two pages...

Well, I won't beat about the bush, this stuff is pretty tricky... nearly as hard as a cross between Mr T, Hulk Hogan and Arnie. But, with a little patience and perseverance (and plenty of [chocolate] [coffee] [marshmallows] — delete as you wish), you'll get there. Once you've got these pages straight in your head, the next ones will be easier to understand.

Photosynthesis

*Right, pen at the ready. Check. Brain switched on. Check. Cuppa piping hot. Check. Sweets on standby. Check.
Okay, I think you're all sorted to start photosynthesis. Finally, take a deep breath and here we go...*

Photosynthesis Takes Place in the Chloroplasts of Plant Cells

1) **Chloroplasts** are **small, flattened organelles** found in **plant cells**.

2) They have a **double membrane** called the **chloroplast envelope**.

3) **Thylakoids** (fluid-filled sacs) are **stacked up** in the chloroplast into structures called **grana** (singular = **granum**). The grana are **linked** together by bits of thylakoid membrane called **lamellae** (singular = **lamella**).

4) Chloroplasts contain **photosynthetic pigments** (e.g. **chlorophyll a**, **chlorophyll b** and **carotene**). These are **coloured substances** that **absorb** the **light energy** needed for photosynthesis. The pigments are found in the **thylakoid membranes** — they're attached to **proteins**. The protein and pigment is called a **photosystem**.

5) There are **two** photosystems used by plants to capture light energy. **Photosystem I** (or **PSI**) absorbs light best at a wavelength of **700 nm** and **photosystem II** (**PSII**) absorbs light best at **680 nm**.

6) Contained within the inner membrane of the chloroplast and **surrounding** the thylakoids is a gel-like substance called the **stroma**. It contains **enzymes**, **sugars** and **organic acids**.

7) Carbohydrates produced by photosynthesis and not used straight away are stored as **starch grains** in the **stroma**.

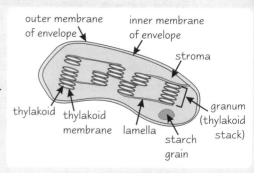

outer membrane of envelope / inner membrane of envelope / stroma / thylakoid / thylakoid membrane / lamella / starch grain / granum (thylakoid stack)

Photosynthesis can be Split into Two Stages

There are actually **two stages** that make up **photosynthesis**:

See p. 18 for loads more information on the Calvin cycle.

① The Light-Dependent Reaction

1) As the name suggests, this reaction **needs light energy**.

2) It takes place in the **thylakoid membranes** of the chloroplasts.

3) Here, light energy is absorbed by **photosynthetic pigments** in the **photosystems** and converted to **chemical energy**.

4) The light energy is used to add a phosphate group to ADP to form **ATP**, and to reduce NADP to form **reduced NADP**. ATP **transfers energy** and reduced **NADP transfers hydrogen** to the light-independent reaction.

5) During the process H_2O is **oxidised** to O_2.

② The Light-Independent Reaction

1) This is also called the **Calvin cycle** and as the name suggests it **doesn't use light energy** directly. (But it does **rely** on the **products** of the light-dependent reaction.)

2) It takes place in the **stroma** of the chloroplast.

3) Here, the **ATP** and **reduced NADP** from the light-dependent reaction supply the **energy** and **hydrogen** to make **glucose** from CO_2.

This diagram shows how the two reactions link together in the chloroplast:

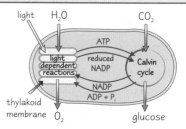

light / H_2O / CO_2 / ATP / light dependent reactions / reduced NADP / Calvin cycle / NADP / $ADP + P_i$ / thylakoid membrane / O_2 / glucose

In the Light-Dependent Reaction ATP is Made by Photophosphorylation

In the light-dependent reaction, the **light energy** absorbed by the photosystems is used for **three** things:

1) Making **ATP** from **ADP** and **inorganic phosphate**. This reaction is called **photophosphorylation** (see p. 15).

2) Making **reduced NADP** from **NADP**.

3) Splitting **water** into **protons** (H^+ ions), **electrons** and **oxygen**. This is called **photolysis** (see p. 15).

The light-dependent reaction actually includes **two types** of **photophosphorylation** — **non-cyclic** and **cyclic**. Each of these processes has **different products**.

Photosynthesis

Non-cyclic Photophosphorylation Produces ATP, Reduced NADP and O_2

To understand the process you need to know that the photosystems (in the thylakoid membranes) are **linked** by **electron carriers**. Electron carriers are **proteins** that **transfer electrons**. The photosystems and electron carriers form an **electron transport chain** — a **chain** of **proteins** through which **excited electrons flow**. All the processes in the diagrams are happening together — I've just split them up to make it easier to understand.

1) Light energy excites electrons in chlorophyll

- **Light energy** is absorbed by **PSII**.
- The light energy **excites electrons** in **chlorophyll**.
- The electrons move to a **higher energy level** (i.e. they have more energy).
- These high-energy electrons **move along** the **electron transport chain** to **PSI**.

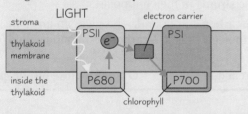

2) Photolysis of water produces protons (H^+ ions), electrons and O_2

- As the excited electrons **from chlorophyll leave PSII** to **move along** the electron transport chain, they must be **replaced**.
- **Light** energy splits **water** into **protons** (H^+ ions), **electrons** and **oxygen**. (So the O_2 in photosynthesis comes from water.)
- The reaction is: $H_2O \longrightarrow 2H^+ + \frac{1}{2}O_2$

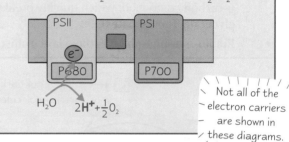

Not all of the electron carriers are shown in these diagrams.

3) Energy from the excited electrons makes ATP...

- The excited electrons **lose energy** as they **move along** the **electron transport chain**.
- This energy is used to **transport protons into** the **thylakoid** so that the thylakoid has a **higher concentration** of protons than the stroma. This forms a **proton gradient** across the membrane.
- Protons move **down** their concentration gradient, into the stroma, **via** an enzyme called **ATP synthase**. The energy from this movement combines **ADP** and **inorganic phosphate** (P_i) to form **ATP**.

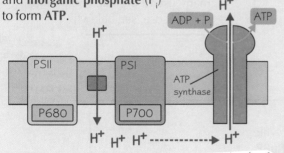

Chemiosmosis is the name of the process where the movement of H^+ ions across a membrane generates ATP. This process also occurs in respiration (see p. 25).

4) ...and generates reduced NADP.

- Light energy is **absorbed** by PSI, which excites the electrons again to an **even higher** energy level.
- Finally, the electrons are **transferred** to **NADP**, along with a **proton** (H^+ ion) from the **stroma**, to form **reduced NADP**.

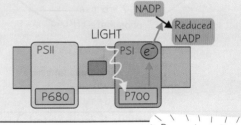

Remember a 'proton' is just another word for a hydrogen ion (H^+).

Cyclic Photophosphorylation Only Produces ATP

Cyclic photophosphorylation **only uses PSI**. It's called 'cyclic' because the electrons from the chlorophyll molecule **aren't** passed onto NADP, but are **passed back** to PSI via electron carriers. This means the electrons are **recycled** and can repeatedly flow through PSI. This process doesn't produce any reduced NADP or O_2 — it **only produces** small amounts of **ATP**.

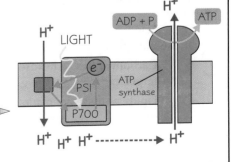

Photosynthesis

Don't worry, you're over the worst of photosynthesis now. Instead of electrons flying around, there's a nice cycle of reactions to learn. What more could you want from life? Money, fast cars and nice clothes have nothing on this...

The **Light-Independent** Reaction is also called the **Calvin Cycle**

1) The Calvin cycle takes place in the **stroma** of the chloroplasts.

2) It makes a molecule called **triose phosphate** from CO_2 and **ribulose bisphosphate** (a 5-carbon compound). Triose phosphate can be used to make **glucose** and other **useful organic substances** (see below).

3) There are a few steps in the cycle, and it needs **ATP** and **H⁺ ions** to keep it going.

4) The reactions are linked in a **cycle**, which means the starting compound, **ribulose bisphosphate**, is **regenerated**.

The Calvin cycle is also called carbon fixation, because carbon from CO_2 is 'fixed' into an organic molecule.

Here's what happens at each stage in the cycle:

> **1** **Carbon dioxide is combined with ribulose bisphosphate to form two molecules of glycerate 3-phosphate**
>
> - CO_2 enters the leaf through the **stomata** and diffuses into the **stroma** of the chloroplast.
> - Here, it's combined with **ribulose bisphosphate (RuBP)**, a **5-carbon** compound. This gives an **unstable 6-carbon** compound, which quickly breaks down into **two molecules** of a **3-carbon** compound called **glycerate 3-phosphate (GP)**.
> - **Ribulose bisphosphate carboxylase (rubisco)** catalyses the reaction between CO_2 and **ribulose bisphosphate**.

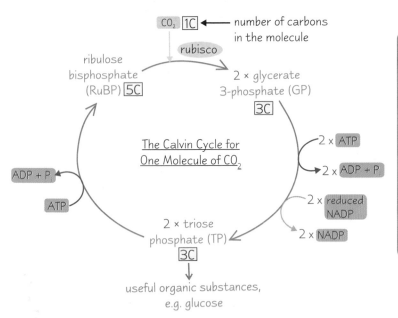

number of carbons in the molecule

The Calvin Cycle for One Molecule of CO_2

useful organic substances, e.g. glucose

> **2** **ATP and reduced NADP are required for the reduction of GP to triose phosphate**
>
> - Now **ATP**, from the **light-dependent** reaction, **provides energy** to turn the **3-carbon** compound, **GP**, into a **different** 3-carbon compound called **triose phosphate (TP)**.
> - This reaction also requires **H⁺ ions**, which come from **reduced NADP** (also from the **light-dependent** reaction). Reduced NADP is **recycled** to **NADP**.
> - **Triose phosphate** is then converted into many **useful organic compounds**, e.g. glucose (see below).

Reduced NADP reduces GP to TP — reduction reactions are explained on p. 15.

> **3** **Ribulose bisphosphate is regenerated**
>
> - **Five** out of every **six** molecules of **TP** produced in the cycle aren't used to make hexose sugars, but to **regenerate RuBP**.
> - Regenerating RuBP uses the **rest** of the ATP produced by the **light-dependent reaction**.

TP and **GP** are **Converted** into **Useful Organic Substances** like **Glucose**

The Calvin cycle is the starting point for making **all** the organic substances a plant needs.
Triose phosphate (TP) and **glycerate 3-phosphate** (GP) molecules are used to make **carbohydrates**, **lipids** and **proteins**:

- **Carbohydrates** — **hexose sugars** (e.g. glucose) are made by joining **two triose phosphate molecules** together and **larger** carbohydrates (e.g. sucrose, starch, cellulose) are made by joining **hexose sugars** together in **different ways**.

- **Lipids** — these are made using **glycerol**, which is synthesised from **triose phosphate**, and **fatty acids**, which are synthesised from **glycerate 3-phosphate**.

- **Proteins** — some **amino acids** are made from **glycerate 3-phosphate**, which are joined together to make proteins.

Photosynthesis

The Calvin Cycle Needs to Turn Six Times to Make One Hexose Sugar

Here's the reason why:

1) **Three turns** of the cycle produces **six** molecules of **triose phosphate** (TP), because two molecules of TP are made for every one CO_2 molecule used.

2) **Five** out of **six** of these TP molecules are used to **regenerate ribulose bisphosphate** (RuBP).

3) This means that for **three turns** of the cycle only **one TP** is produced that's used to make a **hexose sugar**.

4) A hexose sugar has **six carbons** though, so **two TP** molecules are needed to form one hexose sugar.

5) This means the cycle must turn **six times** to produce **two molecules** of **TP** that can be used to make **one hexose sugar**.

6) Six turns of the cycle need **18 ATP** and **12 reduced NADP** from the light-dependent reaction.

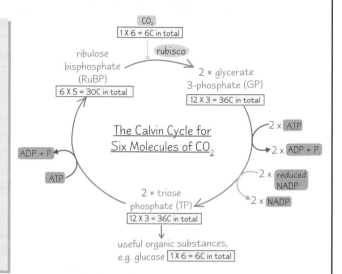

This might seem a bit inefficient, but it keeps the cycle going and makes sure there's always **enough RuBP** ready to combine with CO_2 taken in from the atmosphere.

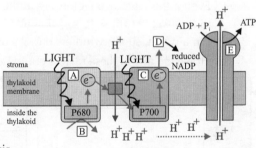

Morag had to turn one million times to make a sock... two million for a scarf.

Practice Questions

Q1 Name two photosynthetic pigments in the chloroplasts of plants.

Q2 At what wavelength does photosystem I absorb light best?

Q3 What three substances does non-cyclic photophosphorylation produce?

Q4 Which photosystem is involved in cyclic photophosphorylation?

Q5 Where in the chloroplasts does the light-independent reaction occur?

Q6 How many carbon atoms are there in a molecule of TP?

Q7 Name two organic substances made from triose phosphate.

Q8 How many CO_2 molecules need to enter the Calvin cycle to make one hexose sugar?

Exam Questions

Q1 The diagram above shows the light-dependent reaction of photosynthesis.
   a) Where precisely in a plant does the light-dependent reaction of photosynthesis occur? [1 mark]
   b) What is A? [1 mark]
   c) Describe process B and explain its purpose. [4 marks]
   d) What is reactant D? [1 mark]

Q2 Rubisco is an enzyme that catalyses the first reaction of the Calvin cycle.
   CA1P is an inhibitor of rubisco.
   a) Describe how triose phosphate is produced in the Calvin cycle. [6 marks]
   b) Briefly explain how ribulose bisphosphate (RuBP) is regenerated in the Calvin cycle. [2 marks]
   c) Explain the effect that CA1P would have on glucose production. [3 marks]

Calvin cycles — bikes made by people that normally make pants...

Next thing we know there'll be male models swanning about in their pants riding highly fashionable bikes. Sounds awful I know, but let's face it, anything would look better than cycling shorts. Anyway, it would be a good idea to go over these pages a couple of times — you might not feel as if you can fit any more information in your head, but you can, I promise.

Limiting Factors in Photosynthesis

I'd love to tell you that you'd finished photosynthesis... but I'd be lying.

There are **Optimum Conditions** for Photosynthesis

The **ideal conditions** for photosynthesis vary from one plant species to another, but the conditions below would be ideal for **most** plant species in temperate climates like the UK.

1. High light intensity of a certain **wavelength**

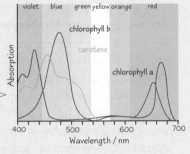

- Light is needed to provide the **energy** for the **light-dependent reaction** — the **higher** the **intensity** of the light, the **more energy** it provides.
- Only certain **wavelengths** of light are used for photosynthesis. The photosynthetic pigments chlorophyll a, chlorophyll b and carotene only **absorb** the **red** and **blue** light in sunlight. (**Green** light is **reflected**, which is why plants look green.)

2. Temperature around 25 °C

- Photosynthesis involves **enzymes** (e.g. ATP synthase, rubisco). If the temperature falls **below 10 °C** the enzymes become **inactive**, but if the temperature is **more than 45 °C** they may start to **denature**.
- Also, at **high** temperatures stomata **close** to avoid losing too much water. This causes photosynthesis to slow down because **less CO_2** enters the leaf when the stomata are closed.

3. Carbon dioxide at 0.4%

- Carbon dioxide makes up **0.04%** of the gases in the atmosphere.
- Increasing this to **0.4%** gives a **higher rate** of photosynthesis, but any higher and the stomata start to **close**.

Plants also need a **constant supply** of water — **too little** and photosynthesis has to **stop** but **too much** and the soil becomes **waterlogged** (**reducing** the uptake of **minerals** such as **magnesium**, which is needed to make **chlorophyll a**).

Light, **Temperature** and **CO_2** can all **Limit Photosynthesis**

1) **All three** of these things need to be at the **right level** to allow a plant to photosynthesise as quickly as possible.

2) If any **one** of these factors is **too low** or **too high**, it will **limit photosynthesis** (slow it down). Even if the other two factors are at the perfect level, it won't make **any difference** to the speed of photosynthesis as long as that factor is at the wrong level.

3) On a warm, sunny, windless day, it's usually **CO_2** that's the limiting factor, and at night it's the **light intensity**.

4) However, **any** of these factors could become the limiting factor, depending on the **environmental conditions**.

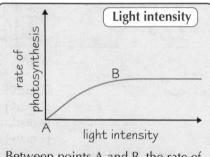

Between points A and B, the rate of photosynthesis is limited by the **light intensity**. So as the light intensity **increases**, so can the rate of photosynthesis. Point B is the **saturation point** — increasing light intensity after this point makes no difference, because **something else** has become the limiting factor. The graph now **levels off**.

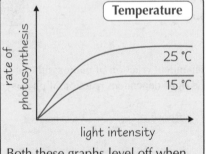

Both these graphs level off when **light intensity** is no longer the limiting factor. The graph at **25 °C** levels off at a **higher point** than the one at **15 °C**, showing that **temperature** must have been a limiting factor at **15 °C**.

The saturation point is where a factor is no longer limiting the reaction — something else has become the limiting factor.

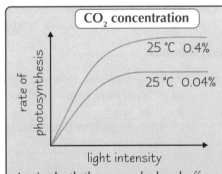

Again, both these graphs level off when **light intensity** is no longer the limiting factor. The graph at **0.4% CO_2** levels off at a **higher point** than the one at **0.04%**, so **CO_2 concentration** must have been a limiting factor at **0.04% CO_2**. The limiting factor here **isn't temperature** because it's the **same** for both graphs (25 °C).

Limiting Factors in Photosynthesis

Growers Use Information About Limiting Factors to Increase Plant Growth

Commercial growers (e.g. farmers) know the **factors** that **limit photosynthesis** and therefore limit **plant growth**. This means they can create an **environment** where plants get the **right amount** of everything that they need, which **increases growth** and so **increases yield**. Growers create optimum conditions in **glasshouses**, in the following ways:

Limiting Factor	Management in Glasshouse
Carbon dioxide concentration	CO_2 is added to the air, e.g. by burning a small amount of propane in a CO_2 generator.
Light	Light can get in through the glass. Lamps provide light at night-time.
Temperature	Glasshouses trap heat energy from sunlight, which warms the air. Heaters and cooling systems can also be used to keep a constant optimum temperature, and air circulation systems make sure the temperature is even throughout the glasshouse.

You Need to be Able to Interpret Data on Limiting Factors

Here are some **examples** of the kind of **data** you might get in the exam:

The graph on the **right** shows the effect on plant growth of **adding carbon dioxide** to a greenhouse.

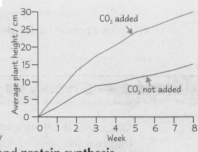

1) In the greenhouse **with added CO_2** plant **growth** was **faster** (the line is steeper) and on average the plants were **larger** after 8 weeks than they were in the control greenhouse (30 cm compared to only 15 cm in the greenhouse where no CO_2 was added).

2) This is because the plants use CO_2 to produce **glucose** by photosynthesis. The more CO_2 they have, the more glucose they can produce, meaning they can **respire more** and so have **more ATP** for **DNA replication**, **cell division** and **protein synthesis**.

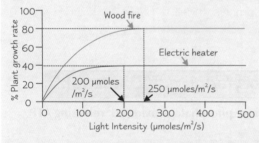

The graph on the **left** shows the effect of **light intensity** on plant growth, and the effect of two **different types** of **heater**.

1) At the start of the graph, the **greater** the **light intensity** the **greater** the **plant growth**.

2) At **200 μmoles/m²/s** of light the **bottom** graph flattens out, showing that CO_2 concentration or temperature is **limiting growth** in these plants.

3) At **250 μmoles/m²/s** of light the **top** graph flattens out.

The difference between the two graphs could be because the **wood fire increases the temperature more** than the electric heater or because it's **increasing the concentration** of CO_2 in the air (an electric heater **doesn't release CO_2**).

Practice Questions

Q1 Name two factors that can limit plant growth.

Q2 How do commercial growers create an optimum level of CO_2 in a glasshouse?

Exam Question

Crop	Yield in glasshouse / kg	Yield grown outdoors / kg
Tomato	1000	200
Lettuce	750	230
Potato	850	680
Wheat	780	550

Q1 The table above shows the yields of various crops when they are grown in glasshouses and when grown outdoors.
   a) Yields are usually higher overall in glasshouses.
      Describe four ways in which conditions can be controlled in glasshouses to increase yields. [4 marks]
   b) Glasshouses are not always financially viable for all crops.
      Which crop above benefits the least from being grown in glasshouses? Explain your answer. [2 marks]

I'm a whizz at the factors that limit revision...

... watching Hollyoaks, making tea, watching EastEnders, walking the dog... not to mention staring into space (one of my favourites). Anyway, an interpreting data question could well come up in the exam — it could be any kind of data, but don't panic if it's not like the graphs above — as long as you understand limiting factors you'll be able to interpret it.

Respiration

From the last gazillion pages you know that plants make their own glucose. Unfortunately, that means now you need to learn how plant and animal cells release energy from glucose. It's not the easiest thing in the world to understand, but it'll make sense once you've gone through it a couple of times.

There are **Four Stages** in **Aerobic Respiration**

1) The four stages in aerobic respiration are **glycolysis**, the **link reaction**, the **Krebs cycle** and **oxidative phosphorylation**.

2) The **first three** stages are a **series of reactions**. The **products** from these reactions are **used** in the **final stage** to produce loads of ATP.

3) The **first** stage happens in the **cytoplasm** of cells and the **other three** stages take place in the **mitochondria**. You might want to refresh your memory of mitochondrion structure before you start.

4) **Anaerobic** respiration **doesn't involve** the **link reaction**, the **Krebs cycle** or **oxidative phosphorylation**. The **products of glycolysis** are converted to ethanol or lactate instead (see the next page for more).

5) All cells use **glucose** to **respire**, but organisms can also **break down** other **complex organic molecules** (e.g. fatty acids, amino acids), which can then be respired.

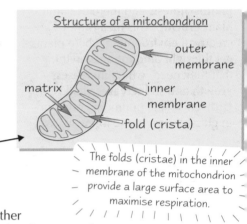

Structure of a mitochondrion

outer membrane
matrix
inner membrane
fold (crista)

The folds (cristae) in the inner membrane of the mitochondrion provide a large surface area to maximise respiration.

Stage 1 — *Glycolysis* Makes *Pyruvate* from *Glucose*

1) Glycolysis involves splitting **one molecule** of glucose (with 6 carbons — 6C) into **two** smaller molecules of **pyruvate** (3C).

2) The process happens in the **cytoplasm** of cells.

3) Glycolysis is the **first stage** of both aerobic and anaerobic respiration and **doesn't need** oxygen to take place — so it's an **anaerobic** process.

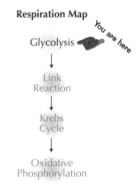

Respiration Map

Glycolysis ← You are here

Link Reaction

Krebs Cycle

Oxidative Phosphorylation

There are **Two Stages** in Glycolysis — *Phosphorylation* and *Oxidation*

First, **ATP** is **used** to **phosphorylate glucose** to triose phosphate. Then **triose phosphate** is **oxidised**, **releasing ATP**. Overall there's a **net gain** of **2 ATP**.

(1) Stage One — Phosphorylation

1) Glucose is **phosphorylated** by adding **2 phosphates** from **2 molecules** of **ATP**.

2) This creates **2 molecules** of **triose phosphate** and **2 molecules** of **ADP**.

(2) Stage Two — Oxidation

1) Triose phosphate is **oxidised** (loses hydrogen), forming **2 molecules** of **pyruvate**.

2) **NAD** collects the hydrogen ions, forming **2 reduced NAD**.

3) **4 ATP** are **produced**, but 2 were used up in stage one, so there's a **net gain** of **2 ATP**.

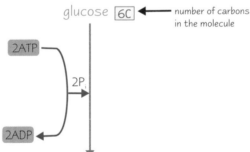

glucose 6C ← number of carbons in the molecule

2ATP

2P$_i$

2ADP

2 × triose phosphate 3C

4ADP + 4P$_i$

2H⁺

2NAD

4ATP

2 reduced NAD

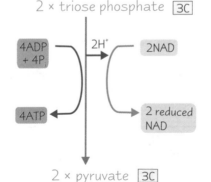

2 × pyruvate 3C

You're probably wondering what now happens to all the products of glycolysis...

1) The **two** molecules of **reduced NAD** go to the **last stage** (oxidative phosphorylation — see page 24).

2) The **two pyruvate** molecules go into the **matrix** of the **mitochondria** for the **link reaction** (see the next page).

Respiration

Stage 2 — the **Link Reaction** converts **Pyruvate** to **Acetyl Coenzyme A**

1) **Pyruvate** is **decarboxylated** — **one carbon atom** is **removed** from pyruvate in the form of **CO₂**.

2) **NAD** is **reduced** — it collects **hydrogen** from pyruvate, changing pyruvate into **acetate**.

3) **Acetate** is combined with **coenzyme A** (CoA) to form **acetyl coenzyme A** (acetyl CoA).

4) **No ATP** is produced in this reaction.

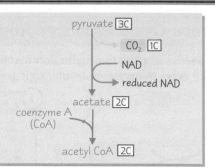

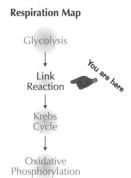

Respiration Map

Glycolysis

Link Reaction

You are here

Krebs Cycle

Oxidative Phosphorylation

The **Link Reaction** occurs **Twice** for every **Glucose Molecule**

Two **pyruvate** molecules are made for **every glucose molecule** that enters glycolysis. This means the **link reaction** and the third stage (the **Krebs cycle**) happen **twice** for every glucose molecule. So for each glucose molecule:

- Two molecules of **acetyl coenzyme A** go into the Krebs cycle (see the next page).
- Two **CO₂** molecules are released as a waste product of respiration.
- **Two** molecules of **reduced NAD** are formed and go to the last stage (oxidative phosphorylation, see page 24).

In **Anaerobic Respiration Pyruvate** is Converted to **Ethanol** or **Lactate**

1) In **aerobic** respiration (where there's **lots** of oxygen) **pyruvate** goes on to the third stage of respiration, the **Krebs cycle** (via the link reaction). In the Krebs cycle, **more ATP** is made and **NAD** is **reduced** (see next page).

2) However, in **anaerobic** respiration (where there's **no** oxygen) **pyruvate** is **converted** into **ethanol** (in **plants** and **yeast**) or **lactate** (in **animal** cells and some **bacteria**):

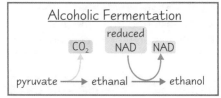

Alcoholic Fermentation

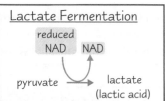

Lactate Fermentation

The production of lactate or ethanol **regenerates NAD**. This means **glycolysis** can **continue** even when there **isn't** much oxygen around, so a **small amount of ATP** can still be **produced** to keep some biological process going... clever.

Practice Questions

Q1   Where in the cell does glycolysis occur?
Q2   Is glycolysis an anaerobic or aerobic process?
Q3   How many ATP molecules are used up in glycolysis?
Q4   What is the product of the link reaction?

Exam Questions

Q1   Describe how a 6-carbon molecule of glucose is converted to pyruvate.   [6 marks]

Q2   At the end of a 100 m sprint runners will have built up lactate in their muscle cells.
      a) Write down an equation to show how lactate is produced.   [2 marks]
      b) What is the advantage of producing lactate in anaerobic respiration?   [2 marks]

No ATP was harmed during this reaction...

Ahhhh... too many reactions. I'm sure your head hurts now, 'cause mine certainly does. Just think of revision as like doing exercise — it can be a pain while you're doing it (and maybe afterwards too), but it's worth it for the well-toned brain you'll have. Just keep going over and over it, until you get the first two stages of respiration straight in your head. Then relax.

Respiration

As you've seen, glycolysis produces a net gain of two ATP. Pah, we can do better than that.
The Krebs cycle and oxidative phosphorylation are where it all happens — ATP galore.

Stage 3 — the **Krebs Cycle** Produces **Reduced Coenzymes** and **ATP**

The Krebs cycle involves a series of **oxidation-reduction reactions**, which take place in the **matrix** of the **mitochondria**. The cycle happens **once** for **every pyruvate** molecule, so it goes round **twice** for **every glucose** molecule.

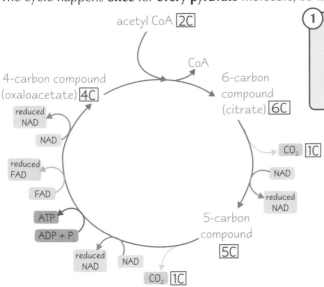

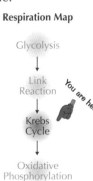

Respiration Map

Glycolysis

↓

Link Reaction

You are here

Krebs Cycle

↓

Oxidative Phosphorylation

1
- **Acetyl CoA** from the link reaction combines with **oxaloacetate** to form **citrate**.
- **Coenzyme A** goes back to the **link reaction** to be used again.

2
- The **6C citrate** molecule is converted to a **5C molecule**.
- **Decarboxylation** occurs, where **CO_2** is **removed**.
- **Dehydrogenation** also occurs, where **hydrogen** is **removed**.
- The hydrogen is used to **produce reduced NAD** from NAD.

3
- The **5C molecule** is then converted to a **4C molecule**. (There are some intermediate compounds formed during this conversion, but you don't need to know about them.)
- **Decarboxylation** and **dehydrogenation** occur, producing **one** molecule of **reduced FAD** and **two** of **reduced NAD**.
- **ATP** is **produced** by the **direct transfer** of a **phosphate** group from an **intermediate** compound to **ADP**. When a phosphate group is directly transferred from one molecule to another it's called **substrate-level phosphorylation**. **Citrate** has now been **converted** into **oxaloacetate**.

Some **Products** of the **Krebs Cycle** are Used in **Oxidative Phosphorylation**

Some products are **reused**, some are **released** and others are used for the **next stage** of respiration:

Product from one Krebs cycle	Where it goes
1 coenzyme A	Reused in the next link reaction
Oxaloacetate	Regenerated for use in the next Krebs cycle
2 CO_2	Released as a waste product
1 ATP	Used for energy
3 reduced NAD	To oxidative phosphorylation
1 reduced FAD	To oxidative phosphorylation

Mr Krebs

Talking about oxidative phosphorylation was always a big hit with the ladies...

Stage 4 — **Oxidative Phosphorylation** Produces **Lots** of **ATP**

1) Oxidative phosphorylation is the process where the **energy** carried by **electrons**, from **reduced coenzymes** (reduced NAD and reduced FAD), is used to **make ATP**. (The whole point of the previous stages is to make reduced NAD and reduced FAD for the final stage.)

2) Oxidative phosphorylation involves two processes — the **electron transport chain** and **chemiosmosis** (see the next page).

Respiration Map

Glycolysis

↓

Link Reaction

↓

Krebs Cycle

↓

Oxidative Phosphorylation

You are here

Respiration

Protons are Pumped Across the Inner Mitochondrial Membrane

So now on to how **oxidative phosphorylation** actually **works**:

1) **Hydrogen atoms** are released from **reduced NAD** and **reduced FAD** as they're oxidised to NAD and FAD. The H atoms **split** into **protons (H⁺)** and **electrons (e⁻)**.

2) The **electrons** move along the **electron transport chain** (made up of three **electron carriers**), **losing energy** at each carrier.

3) This energy is used by the electron carriers to **pump protons** from the **mitochondrial matrix into** the **intermembrane space** (the space **between** the inner and outer **mitochondrial membranes**).

4) The **concentration** of **protons** is now **higher** in the **intermembrane space** than in the mitochondrial matrix — this forms an **electrochemical gradient** (a **concentration gradient** of **ions**).

5) Protons **move down** the **electrochemical gradient**, back into the mitochondrial matrix, via **ATP synthase**. This **movement** drives the synthesis of **ATP** from **ADP** and **inorganic phosphate** (P_i).

6) The movement of H⁺ ions across a membrane, which generates ATP, is called **chemiosmosis**.

7) In the mitochondrial matrix, at the end of the transport chain, the **protons**, **electrons** and **O₂** (from the blood) combine to form **water**. Oxygen is said to be the final **electron acceptor**.

The regenerated coenzymes are reused in the Krebs cycle.

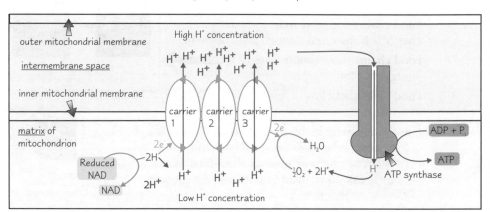

32 ATP Can be Made from One Glucose Molecule

As you know, **oxidative phosphorylation** makes **ATP** using energy from the reduced coenzymes — **2.5 ATP** are made from each **reduced NAD** and **1.5 ATP** are made from each **reduced FAD**. The table on the right shows **how much** ATP a cell can make from **one molecule** of **glucose** in **aerobic respiration**. (Remember, one molecule of glucose produces 2 pyruvate, so the link reaction and Krebs cycle happen twice.)

Stage of respiration	Molecules produced	Number of ATP molecules
Glycolysis	2 ATP	2
Glycolysis	2 reduced NAD	2 × 2.5 = 5
Link Reaction (×2)	2 reduced NAD	2 × 2.5 = 5
Krebs cycle (×2)	2 ATP	2
Krebs cycle (×2)	6 reduced NAD	6 × 2.5 = 15
Krebs cycle (×2)	2 reduced FAD	3 × 1.5 = 3
		Total ATP = 32

The number of ATP produced per reduced NAD or reduced FAD was thought to be 3 and 2, but new research has shown that the figures are nearer 2.5 and 1.5.

Practice Questions

Q1 Where in the cell does the Krebs cycle occur?

Q2 How many times does decarboxylation happen during one turn of the Krebs cycle?

Q3 What do the electrons lose as they move along the electron transport chain in oxidative phosphorylation?

Exam Question

Q1 Carbon monoxide inhibits the final electron carrier in the electron transport chain.
a) Explain how this affects ATP production via the electron transport chain. [2 marks]
b) Explain how this affects ATP production via the Krebs cycle. [2 marks]

The electron transport chain isn't just a FAD with the examiners...

Oh my gosh, I didn't think it could get any worse... You may be wondering how to learn these pages of crazy chemistry, but basically you have to put in the time and go over and over it. Don't worry though, it WILL pay off, and before you know it you'll be set for the exam. And once you know this section you'll be able to do anything, e.g. world domination...

Energy Transfer and Productivity

Some organisms get their energy from the sun and some get it from other organisms, and it's all very friendly. Yeah right.

Energy is Transferred Through Ecosystems

1) An **ecosystem** includes all the **organisms** living in a particular area and all the **non-living** (abiotic) conditions.

2) The **main route** by which energy **enters** an ecosystem is **photosynthesis** (e.g. by plants, see p. 14). (Some energy enters sea ecosystems when bacteria respire chemicals from deep sea vents.)

3) During photosynthesis plants **convert sunlight energy** into a form that can be **used** by other organisms — plants are called **producers** (even though they're only converting the energy, not producing it).

4) Energy is **transferred** through the **living organisms** of an ecosystem when organisms **eat** other organisms, e.g. producers are eaten by organisms called **primary consumers**. Primary consumers are then eaten by **secondary consumers** and secondary consumers are eaten by **tertiary consumers**.

5) Each of the stages (e.g. producers, primary consumers) are called **trophic levels**.

6) **Food chains** and **food webs** show how energy is **transferred** through an ecosystem.

7) **Food chains** show **simple lines** of energy transfer.

8) **Food webs** show **lots** of **food chains** in an ecosystem and how they **overlap**.

9) Energy locked up in the things that **can't be eaten** (e.g. bones, faeces) gets recycled back into the ecosystem by microorganisms called **decomposers** — they **break down dead** or **undigested** material.

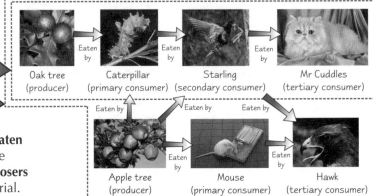

Oak tree (producer) — Eaten by → Caterpillar (primary consumer) — Eaten by → Starling (secondary consumer) — Eaten by → Mr Cuddles (tertiary consumer)

Apple tree (producer) — Eaten by → Mouse (primary consumer) — Eaten by → Hawk (tertiary consumer)

Not All Energy gets Transferred to the Next Trophic Level

1) **Not all** the energy (e.g. from sunlight or food) that's available to the organisms in a trophic level is **transferred** to the **next** trophic level — around **90%** of the **total available energy** is **lost** in various ways.

2) Some of the available energy (**60%**) is **never taken in** by the organisms in the first place. For example:
   - Plants **can't use** all the light energy that reaches their leaves (e.g. some is the **wrong wavelength**).
   - Some **parts** of food, e.g. **roots** or **bones**, **aren't eaten** by organisms so the energy isn't taken in.
   - Some parts of food are **indigestible** so **pass through** organisms and come out as **waste**, e.g. **faeces**.

3) The rest of the available energy (**40%**) is **taken in (absorbed)** — this is called the **gross productivity**. But not all of this is available to the next trophic level either.
   - **30%** of the **total energy** available (75% of the gross productivity) is **lost to the environment** when organisms use energy produced from **respiration** for **movement** or body **heat**. This is called **respiratory loss**.
   - **10%** of the **total energy** available (25% of the gross productivity) becomes **biomass** (e.g. it's **stored** or used for **growth**) — this is called the **net productivity**.

4) **Net productivity** is the amount of energy that's **available** to the **next trophic level**. Here's how it's **calculated:**

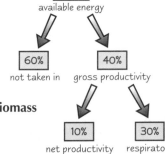

100% available energy

60% not taken in — 40% gross productivity

10% net productivity (available to the next trophic level) — 30% respiratory loss

net productivity = gross productivity – respiratory loss

EXAMPLE: The rabbits in an ecosystem receive **20 000 kJm⁻² yr⁻¹** of energy, but don't take in **12 000 kJm⁻² yr⁻¹** of it, so their gross productivity is **8000 kJm⁻² yr⁻¹** (20 000 – 12 000). They lose **6000 kJm⁻² yr⁻¹** using energy from **respiration**. You can use this to **calculate** the **net productivity** of the rabbits:

net productivity = 8000 – 6000 = 2000 kJm⁻² yr⁻¹

5) You might be asked to **calculate** how **efficient energy transfer** from one trophic level to another is:

The rabbits receive **20 000 kJm⁻² yr⁻¹**, and their **net productivity** is **2000 kJm⁻² yr⁻¹**. So the **percentage efficiency of energy transfer** is:

(2000 ÷ 20 000) × 100 = 10%

Energy Transfer and Productivity

You can also Draw Food Chains as Pyramid Diagrams

1) **Food chains** can be shown by drawing **pyramids** with each block representing a **trophic level**.

2) **Producers** are always on the **bottom**, then **primary consumers** are above them, followed by **secondary consumers** then **tertiary consumers**.

3) The **area** of each block tells you about the **size** of the trophic level.

4) There are **three** types of pyramid — pyramids of **number**, **biomass** and **energy**:

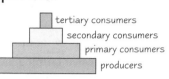

Pyramids of Numbers

- Pyramids of numbers show the **number** of organisms in each trophic level.
- They're not always **pyramid shaped** though — **small numbers** of **big organisms** (like trees) or **large numbers** of **small organisms** (like parasites) change the shape.

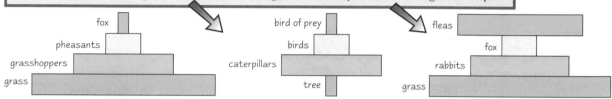

Pyramids of Biomass

- Pyramids of biomass show the **amount** of **biomass** in each trophic level (the **dry mass** of the organisms in kgm^{-2}) at a **single moment** in time.
- They **nearly** always come out pyramid-shaped. An exception is when they're based on **plant plankton** (microorganisms that photosynthesise) — the amount of plant plankton is quite **small** at any **given moment**, but because they have a **short life span** and **reproduce very quickly** there's **a lot** around over a **period of time**.

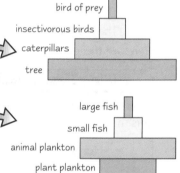

Pyramids of Energy

- Pyramids of energy show the **amount** of **energy** available in each trophic level in **kilojoules** per **square metre** per **year** ($kJm^{-2}yr^{-1}$) — the **net productivity** of each trophic level (see previous page).
- Pyramids of energy are **always** pyramid shaped.

Practice Questions

Q1 State the main way that energy enters an ecosystem.

Q2 What do food webs show?

Q3 What do pyramids of biomass show?

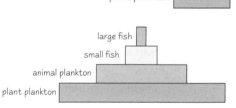

Polar bear — 11 $kJm^{-2}yr^{-1}$
Arctic fox — 137 $kJm^{-2}yr^{-1}$
Arctic hare — 1245 $kJm^{-2}yr^{-1}$
Grass — 13 883 $kJm^{-2}yr^{-1}$

Exam Question

Q1 The pyramid of energy for a food chain is shown above.

a) The respiratory loss of the Arctic hare is 4165 $kJm^{-2} yr^{-1}$.
   Calculate the gross productivity of the Arctic hare, showing your working. [2 marks]

b) Explain why the gross productivity of the Arctic hare is less than the net productivity of the grass. [3 marks]

c) Calculate the percentage efficiency of energy transfer from the Arctic fox to the polar bear. [2 marks]

Boy, do I need an energy transfer this morning...

It's really important to remember that energy transfer through an ecosystem isn't 100% efficient — most gets lost along the way so the next organisms don't get all the energy. Food chains and pyramids are a nice simple way of picturing what happens, but you need to remember that real ecosystems are a bit more complicated, so food webs are needed too.

Farming Practices and Productivity

Farmers may still wear wellies and say ooh-ar, but farming's all about the very serious business of increasing productivity.

Intensive Farming Systems are More Productive than Natural Ecosystems

1) A **natural ecosystem** is an ecosystem that **hasn't been changed** by **human activity**.

2) The **energy input** of a natural ecosystem is the **amount of sunlight** captured by the producers in the ecosystem.

3) **Intensive farming** involves changing an ecosystem by **controlling** the **biotic** or **abiotic conditions**, e.g. the presence of pests or the amount of nutrients available, to make it **more favourable** for crops or livestock.

4) This means intensively farmed **crops** or **livestock** can have **greater net productivity** (a greater **amount** of **biomass**) than **organisms** in **natural ecosystems**.

5) The **energy input** might be **greater** in an intensively farmed area than in a natural ecosystem, e.g. cattle may be given food that's **higher in energy** than their natural food. Or it might be the **same** as a natural ecosystem, e.g. a field of crops still receives the **same** amount of **sunlight** as a natural field.

Intensive Farming Practices Increase Productivity

Intensive farming methods **increase productivity** in different ways:

1) They can **increase** the **efficiency** of **energy conversion** — more of the energy organisms **have** is used for **growth** and less is used for **other activities**, e.g. recovering from **disease** or **movement**.

2) They can remove growth **limiting factors** — **more** of the energy **available** can be used for **growth**.

3) They can **increase energy input** — more energy is **added** to the ecosystem so there's **more energy** for **growth**.

Here are **three** of the main intensive farming practices used:

1) Killing Pest Species

Pests are organisms that **reduce** the **productivity** of **crops** by reducing the amount of energy available for **growth**. This means the crops are **less efficient** at **converting energy**. Here are **three** ways that farmers reduce pest numbers:

Using chemical pesticides

- **Herbicides** kill **weeds** that **compete** with agricultural crops for **energy**. Reducing competition means crops receive **more energy**, so they grow **faster** and become **larger**, **increasing** productivity.
- **Fungicides** kill **fungal infections** that **damage** agricultural crops. The crops **use more** energy for **growth** and **less** for fighting infection, so they grow **faster** and become **larger**, **increasing** productivity.
- **Insecticides** kill **insect** pests that **eat** and **damage** crops. Killing insect pests means **less biomass is lost** from crops, so they grow to be **larger**, which means productivity is **greater**.

Using **chemical pesticides** raises **environmental issues**:

1) They may **directly** affect (**damage** or **kill**) other **non-pest species**, e.g. butterflies.

2) They may **indirectly** affect other **non-pest species**, e.g. eating a lot of **primary consumers** that each contain a **small amount of chemical pesticide** can be enough to **poison a secondary consumer**.

There are also **economic issues**:

Chemical pesticides can be **expensive**. It may not be **profitable** for some farmers to use chemical pesticides — their **cost** may be **greater** than the **extra money** made from **increased productivity**.

Using biological agents

Biological agents reduce the **numbers of pests**, so crops lose **less energy** and **biomass**, **increasing** productivity.

- **Natural predators** introduced to the ecosystem **eat** the pest species, e.g. ladybirds eat greenfly.
- **Parasites** live in or lay their **eggs** on a **pest insect**. Parasites either **kill** the insect or **reduce** its ability to **function**, e.g. some species of wasps lay their eggs inside caterpillars — the eggs hatch and **kill** the caterpillars.
- **Pathogenic** (disease-causing) **bacteria** and **viruses** are used to kill pests, e.g. the bacterium *Bacillus thuringiensis* produces a **toxin** that kills a wide range of **caterpillars**.

Using biological agents raises **environmental issues**:

1) Natural predators introduced to an ecosystem may **become a pest species** themselves.

2) Biological agents can **affect** (damage or kill) other **non-pest species**.

There are also **economic issues**:

Biological agents may be less **cost-effective** than chemical pesticides, i.e. they may increase productivity **less** in the **short term** for the **same amount** of money invested.

Farming Practices and Productivity

Integrated systems use **both chemical pesticides** (e.g. insecticides) and **biological agents** (e.g. parasites).

1) The **combined effect** of using both can reduce pest numbers **even more** than either method **alone**, meaning **productivity** is **increased** even more.

2) Integrated systems can **reduce costs** if one method is **particularly expensive** — the expensive method can be used **less** because the two methods are used **together**.

3) Integrated systems can **reduce** the **environmental impact** of things like pesticides, because **less** is used.

2) Using Fertilisers

Fertilisers are chemicals that provide crops with **minerals** needed **for growth**, e.g. **nitrates**. Crops **use up** minerals in the soil as they **grow**, so their growth is **limited** when there **aren't enough** minerals. Adding fertiliser **replaces** the lost minerals, so **more energy** from the ecosystem can be used to grow, **increasing** the **efficiency** of energy conversion.

1) **Natural** fertilisers are **organic** matter — they include **manure** and **sewage sludge** (that's "muck" to you and me).

2) **Artificial** fertilisers are **inorganic** — they contain **pure chemicals** (e.g. ammonium nitrate) as powders or pellets.

Using fertilisers raises **environmental issues**:

1) Fertiliser can be washed into **rivers** and **ponds**, **killing fish** and **plant life** because of **eutrophication** (see p. 34).

2) Using fertilisers changes the **balance** of **nutrients** in the soil — **too much** of a particular nutrient can cause crops and other plants to **die**.

There are also **economic issues**:

Farmers need to get the **amount** of fertiliser they apply **just right**. **Too much** and money is **wasted** as excess fertiliser is **washed away** (causing **eutrophication**). **Too little** and productivity **won't** be increased, so **less money** can be made from **selling** the crop.

3) Rearing Livestock Intensively

Rearing livestock **intensively** involves **controlling** the **conditions** they live in, so **more** of their **energy** is used for **growth** and **less** is used for **other activities** — the **efficiency** of energy conversion is increased so **more biomass** is produced and productivity is **increased**.

1) Animals may be kept in **warm**, **indoor** pens where their **movement** is **restricted**. **Less energy** is **wasted** keeping **warm** and **moving around**.

2) Animals may be given **feed** that's **higher in energy** than their natural food. This **increases** the **energy input**, so **more energy** is available for **growth**.

The benefits are that **more food** can be produced in a **shorter** space of time, often at **lower cost**. However, enhancing productivity by intensive rearing raises **ethical issues**. For example, some people think the **conditions** intensively reared animals are kept in cause the animals **pain**, **distress** or restricts their **natural behaviour**, so it **shouldn't be done**.

Practice Questions

Q1 Name two types of pesticide.

Q2 What are fertilisers?

Q3 What does intensive farming involve?

Exam Question

Q1 Organic farmers don't use artificial chemicals on their land. Describe and explain how an organic farmer might increase productivity by reducing pest numbers on their farm. [5 marks]

Farming practices — baa-aa-aa-rmy...

Crikey, so farming's not just about getting up early to feed the chooks then — farmers want to produce as much food as they can so some of them use intensive methods. Make sure you pay attention to all the environmental, economic and ethical issues — you need to be able to talk about them to evaluate the consequences of using intensive methods.

The Carbon Cycle and Global Warming

Carbon — found in plants, animals, your petrol tank and on your burnt toast.

The **Carbon Cycle** shows how **Carbon** is **Passed On** and **Recycled**

All organisms need carbon to make **essential compounds**, e.g. plants use CO_2 in photosynthesis to make glucose.
The **carbon cycle** is how carbon **moves** through **living organisms** and the **non-living environment**.
It involves four processes — **photosynthesis**, **respiration**, **decomposition** and **combustion**:

1) **Carbon** (in the form of CO_2 from **air** and **water**) is **absorbed** by plants when they carry out **photosynthesis** — it becomes carbon compounds in **plant tissues**.

2) Carbon is **passed on** to **primary consumers** when they **eat** the plants. It's passed on to **secondary** and **tertiary consumers** when they eat other consumers.

3) All living organisms **die** and the carbon compounds in the **dead organisms** are digested by **microorganisms** called **decomposers**, e.g. bacteria and fungi. Feeding on dead organic matter is called **saprobiontic nutrition**.

4) Carbon is **returned** to the air (and water) as **all living organisms** (including the decomposers) carry out **respiration**, which **produces CO_2**.

5) If dead organic matter ends up in places where there **aren't any** decomposers, e.g. deep oceans or bogs, their carbon compounds can be turned into **fossil fuels** over **millions of years** (by heat and pressure).

6) The carbon in fossil fuels is **released** when they're **burnt** — this is called **combustion**.

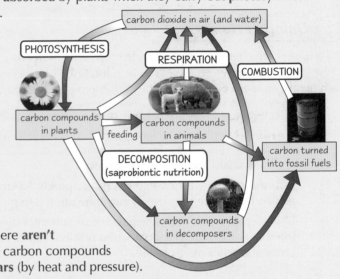

Respiration and **Photosynthesis** Cause Fluctuations in CO_2 **Concentration**

Respiration (which is carried out by **all** organisms) **adds** CO_2 to the atmosphere. **Photosynthesis removes** CO_2 from the atmosphere. The **amount** of respiration and photosynthesis going on **varies** on a **daily** and a **yearly** basis, so the amount of **atmospheric CO_2 changes**.

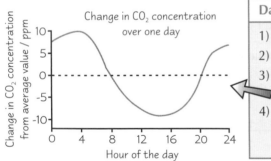

Daily change in CO_2 concentration

1) **Respiration** is carried out **constantly** through the **day and night**.

2) **Photosynthesis** only takes place during the **daylight** hours.

3) CO_2 concentration **falls** during the **day** because it's being **removed** by plants as they carry out photosynthesis.

4) CO_2 concentration **increases** at **night** because it's **no longer** being removed (**no photosynthesis** is happening), but all organisms are **still respiring** and **adding** CO_2 to the atmosphere.

Yearly change in CO_2 concentration

1) Most **plant life** exists in the **northern hemisphere** because that's where **most land** is.

2) Most plant **growth** occurs in the **summer** (June-Aug in the northern hemisphere) because that's when the **light intensity** is greatest — **more photosynthesis** can occur, which means there's **more energy** to grow.

3) CO_2 concentration **falls** during the **summer** because **more** is being **removed** from the atmosphere as **more plants** are photosynthesising.

4) CO_2 concentration **increases** throughout **autumn** and **winter** (Sep-April in the northern hemisphere) because **less** is being **removed** from the atmosphere, as **fewer plants** are **photosynthesising**.

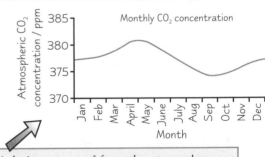

The Carbon Cycle and Global Warming

Global Warming is Caused by Increasing CO₂ and Methane Concentrations

1) **Global warming** is the term for the **increase** in **average global temperature** over the last century.

2) There's a **scientific consensus** that this increase has been **caused** by **human activity**. (It can't be explained by **natural causes**, which happen **more slowly**.)

3) Human activity has caused global warming by **enhancing** the **greenhouse effect** — the effect of greenhouse gases absorbing outgoing **energy**, so that less is **lost** to space.

4) The greenhouse effect is **essential** to keep the planet warm, but **too much** greenhouse gas in the atmosphere means the planet **warms up**.

5) **Two** of the main greenhouse gases are **CO₂** and **methane**.

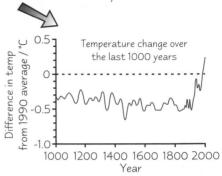

Carbon dioxide (CO₂)

- **Atmospheric CO₂** concentration has **increased rapidly** since the **mid-19th century** from **280 ppm** (parts per million) to nearly **380 ppm**. The concentration had been **stable** for the previous **10 000 years**.

- CO₂ concentration is **increasing** as more **fossil fuels** like coal, oil, natural gas and petrol are **burnt**, e.g. in power stations or in cars. Burning fossil fuels **releases CO₂**.

- CO₂ concentration is also **increased** by the **destruction** of **natural sinks** (things that keep CO₂ **out** of the atmosphere by storing **carbon**). E.g. trees are a big CO₂ sink — they store the carbon as **organic compounds**. CO₂ is **released** when trees are **burnt**, or when **decomposers break down** the organic compounds and **respire**.

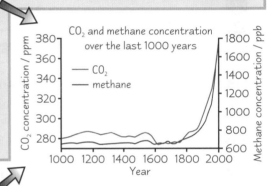

Methane

- **Atmospheric methane** concentration has **increased rapidly** since the **mid-19th century** from **700 ppb** (parts per billion) to **1700 ppb** in **2000**. The level had been **stable** for the previous **850 years**.

- Methane concentration is **increasing** because **more** methane is being **released** into the atmosphere, e.g. because **more fossil fuels** are being **extracted**, there's more **decaying waste** and there are **more cattle** which give off methane as a **waste gas**.

- Methane can also be released from **natural stores**, e.g. **frozen ground** (permafrost). As temperatures **increase** it's thought these stores will **thaw** and release **large amounts** of methane into the atmosphere.

An increase in **human activities** like **burning fossil fuels** (for industry and in cars), **farming** and **deforestation** has **increased** atmospheric concentrations of CO₂ and methane. This has **enhanced** the greenhouse effect and **caused** a rise in average global temperature — **global warming**.

Practice Questions

Q1 What is saprobiontic nutrition?

Q2 Why does CO₂ concentration decrease during daylight hours?

Exam Questions

Q1 Describe how carbon is cycled through living organisms and the non-living environment. [6 marks]

Q2 a) Explain why atmospheric CO₂ and methane concentrations have increased since the mid-19th century. [6 marks]
  b) What is global warming? How are rising atmospheric CO₂ and methane concentrations causing it? [4 marks]

Daily pattern of my concentration — low during lessons and revision...

I know, I know, you might think these pages seem a bit too geographical, but I say it's the other way round — those boring geographers have nicked our biology. The carbon cycle might look a bit messy, but it isn't as complicated as it looks. You just need to break it down into the four processes of photosynthesis, respiration, combustion and decomposition.

Effects of Global Warming

Global warming might mean you can wear a bikini in Scotland, but it's bad news for some organisms...

Global Warming Could Affect all Organisms

Increasing **CO₂ concentration** is causing **global warming**, which is leading to other climate changes, e.g. different **rainfall patterns** and changes to **seasonal weather patterns**. All organisms could be **affected** by this, but **different organisms** could be affected in **different ways**:

Crop yield

The **increasing CO₂ concentration** that's **causing** global warming could **also** be **causing** an **increase** in **crop yields** (the **amount** of crops produced from an area). CO₂ concentration is a **limiting factor** for photosynthesis (see p. 20), so increasing global CO₂ concentration could mean crops grow **faster**, **increasing** crop yields.

Insect pests

1) Climate change may affect the **life cycle** of some insect species. For example, it's thought that increasing global temperature (**global warming**) means some insects go through their **larval stage** quicker and emerge as **adults earlier**, e.g. some butterflies may spend **10** fewer days as larvae for every **1 °C** rise in temperature.

2) Climate change may also affect the **numbers** of some insect species:

- Some species are becoming **more** abundant, e.g. **warmer** and **wetter** summers in some places have led to an **increase** in the number of **mosquitoes**.

- Other species may become **less** abundant, e.g. some **tropical** insect species can only thrive in **specific temperature ranges**, so if it gets **too hot** fewer insects may be able to **reproduce successfully**.

Wild animals and plants

1) Climate change could affect the **distribution** of many wild **animal** and **plant** species:

- Some species may become **more** widely distributed, e.g. species that need **warmer temperatures** may spread **further** as the conditions they **thrive** in exist over a **wider** area.

- Other species may become **less** widely distributed, e.g. species that need **cooler temperatures** may have **smaller** ranges as the conditions they **thrive** in exist over a **smaller** area.

2) Climate change could also affect the **number** of wild animals and plants:

- Some species are becoming **more** abundant, e.g. **boarfish** are increasing in number in parts of the Atlantic Ocean where sea temperature is **rising**.

- Other species are becoming **less** abundant, e.g. **polar bears** need frozen sea ice to hunt and **global warming** is causing more sea ice to **melt**. It's thought that the number of polar bears is **decreasing** because there isn't enough sea ice for them to hunt on.

You Need to be able to Analyse Data on the Effects of Global Warming

Analysing data's pretty important when looking at the **effects** of global warming. Here are a few examples:

1 **Example 1 — Temperature and Crop Yield**

A study was carried out to investigate whether **rising growing season temperature** is affecting **crop yields**. The results of the study are shown on the graph. You might be asked to:

1) **Describe the data:**

The **temperature fluctuated** between **1970** and **2000**, but the general trend was a **steady increase** from just under **17 °C** to just under **18 °C**.

The **wheat yield** also showed a trend of **increasing** from around **1.6 tons** per hectare in **1970** to around **2.7 tons** per hectare in **2000**.

2) **Draw a conclusion:**

There's a **positive correlation** between **temperature** and **wheat yield**. The increasing growing season temperature could be **linked** to the increasing wheat yields.

Even though there's a **correlation**, you can't conclude that the increase in temperature **caused** the increase in wheat yield — there could have been **other factors** involved, e.g. **increased CO₂ concentration**. **Other studies** would need to be carried out to **investigate** the effect of these other factors.

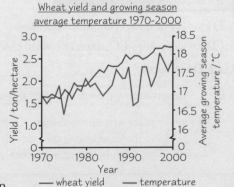

Wheat yield and growing season average temperature 1970-2000

— wheat yield   — temperature

Effects of Global Warming

② Example 2 — *Temperature* and *Insect Numbers*

A study counted the **number** of greenfly in an area from 1960 to 2000.
A separate study collected data on **global temperature** at the same time.
The results are shown on the graph. You might be asked to:

1) **Describe the data:**

   The **temperature fluctuated** between **1960** and **2000**, but the general
   trend was a **steady increase** from just over **13.8 °C** to just over **14.4 °C**.

   The **number of greenfly** also **fluctuated** with a generally **increasing**
   trend from around **110** in **1960** to just around **480** in **2000**.

2) **Draw a conclusion:**

   There's a **positive correlation** between **temperature** and **numbers of greenfly**.
   The increasing global temperature could be **linked to** the increasing greenfly numbers.

3) **Suggest an explanation for your conclusion:**

   Greenfly numbers **could** be increasing because higher temperatures may **increase** their **food supply**, e.g. the rate
   of **photosynthesis** may **increase** at higher temperatures, allowing plants to **grow faster** and **become larger**.

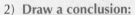

Temperature and number of greenfly caught

③ Example 3 — *Temperature* and the *Distribution* of *Organisms*

A study was carried out to investigate the changing **distribution** of **subtropical plankton** species in the north Atlantic.
The results are shown below, along with data that's been collected on **global sea surface temperature**.
You might be asked to:

1) **Describe the data:**

   Sea surface temperature fluctuated around the
   average between **1950** and **1978**, then there was
   a **steady increase** between 1978 and 2000, up to
   just over **0.3 °C** greater than the average.

   Subtropical plankton species were found in the sea
   south of the UK in 1958-1981. By 2000-2002 their
   distribution had moved **further north** along the west
   coast of the UK and Ireland to the **Arctic Ocean**.

2) **Draw a conclusion:**

   There's a link between **rising global sea surface temperature** and
   the **northward** change in **distribution** of subtropical plankton.

   The data shows a **link**, but you can't say that the increase in temperature **caused**
   the change in distribution — there could have been **other factors** involved,
   e.g. **overfishing** could have removed plankton **predator species**.

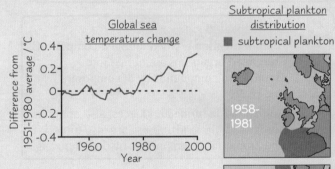

Global sea temperature change

Subtropical plankton distribution

■ subtropical plankton

1958-1981

2000-2002

Practice Questions

Q1 Give one way that climate change is affecting populations of insect pests.

Q2 Give one way that climate change is affecting wild animal species.

Exam Question

Q1 The graph on the right shows CO_2 concentration and corn yield.

    a) Describe what the graph is showing.   [4 marks]

    b) Draw a conclusion.   [1 mark]

    c) Use your knowledge to suggest an explanation for your conclusion.   [2 marks]

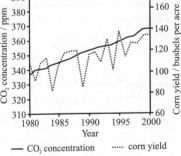

CO_2 concentration and corn yield

— CO_2 concentration ····· corn yield

Global warming effects — not as much fun as special effects...

*Boy, that wasn't fun, but the business of analysing data is an important one if you want to profit in your exam. There could
be lots of questions on data, so have a good read through these examples, and never mix up correlation and cause.*

The Nitrogen Cycle and Eutrophication

Sorry, there's some more cycling to do here — sadly, the nitrogen cycle's a little bit more tiring than the carbon cycle.

The **Nitrogen Cycle** shows how **Nitrogen** is **Passed on** and **Recycled**

Plants and animals **need** nitrogen to make **proteins** and **nucleic acids** (DNA and RNA). The atmosphere's made up of about 78% nitrogen, but plants and animals **can't use it** in that form — they need **bacteria** to **convert** it into **nitrogen compounds** first. The **nitrogen cycle** shows how nitrogen is **converted** into a useable form and then **passed** on between different **living** organisms and the **non-living** environment.

The nitrogen cycle includes **food chains** (nitrogen is passed on when organisms are eaten), and four different processes that involve bacteria — **nitrogen fixation**, **ammonification**, **nitrification** and **denitrification**:

1 **Nitrogen fixation**

- **Nitrogen fixation** is when nitrogen **gas** in the atmosphere is turned into **ammonia** by **bacteria** called *Rhizobium*. The ammonia can then be **used** by plants.
- *Rhizobium* are found inside **root nodules** (growths on the roots) of **leguminous** plants (e.g. peas, beans and clover).
- They form a **mutualistic** relationship with the plants — they provide the plant with **nitrogen compounds** and the plant provides them with **carbohydrates**.

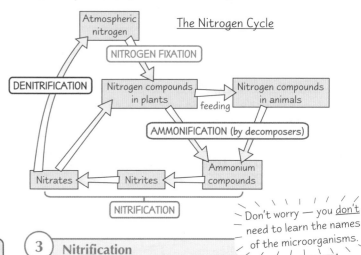

The Nitrogen Cycle

Don't worry — you don't need to learn the names of the microorganisms.

2 **Ammonification**

- **Ammonification** is when nitrogen compounds from **dead organisms** are turned into **ammonium compounds** by **decomposers** (see p. 26).
- Animal **waste** (**urine** and **faeces**) also contains nitrogen compounds. These are also turned into ammonium compounds by decomposers.

3 **Nitrification**

- **Nitrification** is when **ammonium compounds** in the soil are **changed** into **nitrogen compounds** that can then be **used** by plants.
- First **nitrifying bacteria** (e.g. *Nitrosomonas*) change **ammonium compounds** into **nitrites**.
- Then other nitrifying bacteria called *Nitrobacter* change **nitrites** into **nitrates**.

4 **Denitrification**

- **Denitrification** is when nitrates in the soil are **converted** into **nitrogen gas** by **denitrifying bacteria** — they use nitrates in the soil to carry out **respiration** and produce nitrogen gas.
- This happens under **anaerobic conditions** (where there's **no** oxygen), e.g. in **waterlogged** soils.

Parts of the nitrogen cycle can also be carried out **artificially** and on an **industrial** scale. The **Haber process** produces **ammonia** from **atmospheric nitrogen** — it's used to make things like **fertilisers**.

Nitrogen Fertilisers can Leach into Water and Cause Eutrophication

Leaching is when **water-soluble** compounds in the soil are **washed away**, e.g. by rain or irrigation systems. They're often **washed** into nearby **ponds** and **rivers**. If **nitrogen fertiliser** is leached into waterways (e.g. when **too much** is applied to a field) it can cause **eutrophication**:

1) **Nitrates leached** from fertilised fields stimulate the **growth** of **algae** in ponds and rivers.
2) Large amounts of algae **block light** from reaching the plants below.
3) Eventually the **plants die** because they're **unable** to photosynthesise enough.
4) **Bacteria** feed on the dead plant matter.
5) The **increased** numbers of **bacteria reduce** the **oxygen** concentration in the water by carrying out **aerobic respiration**.
6) **Fish** and other aquatic organisms **die** because there **isn't enough dissolved oxygen**.

Hey, who turned out the lights?

The Nitrogen Cycle and Eutrophication

You've got to know how to **analyse data**, so here's an example of the kind of thing you might get in your exam:

A study was conducted to investigate the effect, on a nearby **river**, of adding **fertiliser** to **farmland**.

The **oxygen** and **algal** content of a river that runs past a field where **nitrate fertiliser** had been applied, was measured **at the field** and up to a distance of **180 m** away. A similar **control river** next to an **unfertilised** field was also studied. The results are shown in the graphs on the right.

1) **Describe the data:**

The **algal content** of the water **increases** sharply from **10 000 cells cm^{-3}** at the field to **95 000 cells cm^{-3}** at a distance of **60 m** from the field. Algal content then **decreases** beyond **60 m** to **10 000 cells cm^{-3}** at **180 m**.

The **oxygen content** of the water **decreases** from **8 mgdm^{-3}** at the field to **2 mgdm^{-3}** at a distance of **80 m** from the field. The oxygen content then **increases** beyond **80 m** up to **13 mgdm^{-3}** at 180 m, where it begins to **level off**.

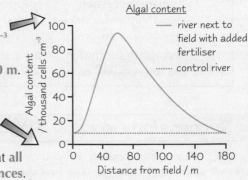

The control river showed a **steady algal content** of **10 000 cells cm^{-3} at all distances**, as well as a **steady oxygen content** of **8 mgdm^{-3} at all distances**.

2) **Draw a conclusion:**

There's a **negative correlation** between the algal content and the oxygen content of the water — as the algal content **increases**, the oxygen content **decreases**, and vice versa.

3) **Evaluate the methodology:**

A control river was used which helps to control the effect of **some variables**, e.g. water temperature. But it doesn't remove the effect of **all** variables, e.g. **different organisms** may live in the control river, which could **affect** the algal or oxygen content.

The experiment only looked at **two rivers**, which means the **sample size** was **small**. Studying other rivers may have produced **different results** and a **different conclusion**. **More experiments** and results would be needed to make the data **more reliable**.

4) **Suggest an explanation for your conclusion:**

The results **suggest leaching** of the fertiliser and **eutrophication** have occurred. **Nitrate fertilisers** from the field could have **leached** into the river and caused the algal content of the river to **increase** by **stimulating** algal growth. The increased algal content could have **prevented light** from reaching plants **below**, causing them to die and be decomposed by **bacteria**. The bacteria **use up** the oxygen in the river when carrying out **aerobic respiration**, resulting in **decreased** dissolved oxygen levels.

Practice Questions

Q1   What is denitrification?
Q2   What is leaching?
Q3   Briefly describe eutrophication.

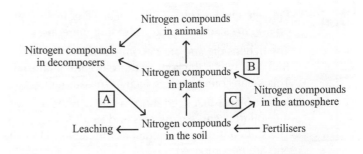

Exam Question

Q1   The diagram on the right shows the nitrogen cycle.

   a) Name the processes labelled A and C in the diagram.   [2 marks]
   b) Name and describe process B in detail.   [3 marks]

Nitrogen fixation — cheaper than a shoe fixation...

The nitrogen cycle's not as bad as it seems. Divide up the four processes of nitrogen fixation, ammonification, nitrification and denitrification and learn them separately, then hey presto — you've learnt the whole cycle. Eutrophication's alright too, it's just got a scary name. Watch out though, they like to combine all this stuff and get you to interpret data in the exam.

Succession

Repeat after me: successful succession involves several simple successive seral stages.

Succession is the Process of Ecosystem Change

Succession is the process by which an **ecosystem** (see p. 4) **changes** over **time**. The **biotic conditions** (e.g. **plant** and **animal communities**) change as the **abiotic conditions** change (e.g. **water** availability). There are **two** types of succession:

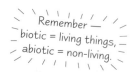

Remember — biotic = living things, abiotic = non-living.

1) **Primary succession** — this happens on land that's been **newly formed** or **exposed**, e.g. where a **volcano** has erupted to form a **new rock surface**, or where **sea level** has **dropped** exposing a new area of land. There's **no soil** or **organic material** to start with, e.g. just bare rock.

2) **Secondary succession** — this happens on land that's been **cleared** of all the **plants**, but where the **soil remains**, e.g. after a **forest fire** or where a forest has been **cut down by humans**.

Succession Occurs in Stages called Seral Stages

1) **Primary succession** starts when species **colonise** a new land surface. **Seeds** and **spores** are blown in by the **wind** and begin to **grow**. The **first species** to colonise the area are called **pioneer species** — this is the **first seral stage**.

   • The **abiotic conditions** are hostile (harsh), e.g. there's no soil to **retain water**. Only pioneer species **grow** because they're **specialised** to cope with the harsh conditions, e.g. **marram grass** can grow on sand dunes near the sea because it has **deep roots** to get water and can **tolerate** the salty environment.

   • The pioneer species **change** the **abiotic conditions** — they **die** and **microorganisms decompose** the dead **organic material** (**humus**). This forms a **basic soil**.

   • This makes conditions **less hostile**, e.g. the basic soil helps to **retain water**, which means **new organisms** can move in and grow. These then die and are decomposed, adding **more** organic material, making the soil **deeper** and **richer in minerals**. This means **larger plants** like **shrubs** can start to grow in the deeper soil, which retains **even more** water.

2) **Secondary succession** happens in the **same way**, but because there's already a **soil layer** succession starts at a **later seral stage** — the pioneer species in secondary succession are **larger plants**, e.g. shrubs.

3) At each stage, **different** plants and animals that are **better adapted** for the improved conditions move in, **out-compete** the plants and animals that are already there, and become the **dominant species** in the ecosystem.

4) As succession goes on, the ecosystem becomes **more complex**. New species move in **alongside** existing species, which means the **species diversity** (the number of **different species** and the **abundance** of each species) **increases**.

5) The **final seral stage** is called the **climax community** — the ecosystem is supporting the **largest** and **most complex** community of plants and animals it can. It **won't change** much more — it's in a **steady state**.

This example shows primary succession on bare rock, but succession also happens on sand dunes, salt marshes and even on lakes.

Example of primary succession — bare rock to woodland

1) **Pioneer species colonise** the rocks. E.g. **lichens** grow **on** and **break down** rocks, **releasing minerals**.

2) The lichens **die** and are **decomposed** helping to form a **thin soil**, which thickens as more **organic material** is formed. This means other species such as **mosses** can **grow**.

3) **Larger plants** that need **more water** can move in as the soil **deepens**, e.g. **grasses** and **small flowering plants**. The soil **continues to deepen** as the larger plants die and are decomposed.

4) **Shrubs**, **ferns** and **small trees** begin to grow, **out-competing** the grasses and smaller plants to become the **dominant** species. **Diversity increases**.

5) Finally, the soil is **deep** and **rich** enough in **nutrients** to support **large trees**. These become the dominant species, and the **climax community** is formed.

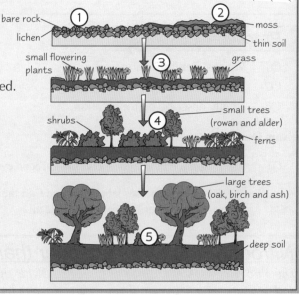

Succession

Different Ecosystems have Different Climax Communities

Which species make up the climax community depends on what the **climate's** like in an ecosystem. The climax community for a **particular** climate is called its **climatic climax**. For example:

- In a **temperate climate** there's **plenty** of **available water**, **mild temperatures** and not much **change** between the seasons. The climatic climax will contain **large trees** because they **can grow** in these conditions once **deep soils** have developed.

- In a **polar climate** there's **not much available water**, temperatures are **low** and there are **massive changes** between the seasons. Large trees **won't ever** be able to grow in these conditions, so the climatic climax contains only **herbs** or **shrubs**, but it's still the **climax community**.

Conservation Often Involves Managing Succession

Human activities can **prevent succession**, stopping a climax community from **developing**. When succession is stopped **artificially** like this the climax community is called a **plagioclimax**. For example:

> A **regularly mown** grassy field **won't develop** shrubs and trees (**woody plants**), even if the climate of the ecosystem could support them. The **growing points** of the woody plants are **cut off** by the lawnmower, so larger plants **can't establish** themselves. The **longer** the interval between mowing, the **further** succession can progress and the more **diversity increases**. But with **more frequent** mowing, succession can't progress and diversity will be **lower** — only the grasses can **survive** being mowed.

Man had been given a mighty weapon with which they would tame the forces of nature.

Conservation (the **protection** and **management** of ecosystems) sometimes involves preventing succession in order to **preserve** an ecosystem in its **current** seral stage. For example, there are large areas of **moorland** in **Scotland** that provide **habitats** for many species of plants and animals. If the moorland was left to **natural processes**, succession would lead to a **climax community** of **spruce forest**. This would mean the **loss** of the moorland habitat and could lead to the loss of some of the plants and animals that **currently** live there. Preventing succession keeps the moorland ecosystem **intact**. There are a couple of ways to **manage succession** to **conserve** the moorland ecosystem:

1) **Animals** are allowed to **graze** on the land. This is similar to **mowing** — the animals eat the **growing points** of the shrubs and trees, which **stops** them from establishing themselves and helps to keep vegetation **low**.

2) **Managed fires** are lit. After the fires, **secondary succession** will occur on the moorland — the species that grow back **first** (**pioneer species**) are the species that are being **conserved**, e.g. heather. Larger species will take **longer** to grow back and will be **removed again** the next time the moor's burnt.

Practice Questions

Q1 What is the difference between primary and secondary succession?

Q2 What is the name given to species that are the first to colonise an area during succession?

Q3 What is meant by a climax community?

Exam Question

Q1 A farmer has a field where he plants crops every year. When the crops are fully grown he removes them all and then ploughs the field (churns up all the plants and soil so the field is left as bare soil). The farmer has decided not to plant crops or plough the field for several years.

a) Describe, in terms of succession, what will happen in the field over time. [6 marks]

b) Explain why succession doesn't usually take place in the farmer's field. [2 marks]

Revision succession — bare brain to a woodland of knowledge...

When answering questions on succession, examiners are pretty keen on you using the right terminology — that means saying "pioneer species" instead of "the first plants to grow there". This stuff's all quite wordy, but the concept of succession is simple enough — some plants start growing, change the environment so it's less hostile, then others can move in.

Conservation

Conservation is important for us and the environment — won't somebody think of the polar bears...

Conserving Species and Habitats is Important for Many Reasons

Conservation is the **protection** and **management** of **species** and **habitats** (**ecosystems**).
It's **important** for **many reasons**:

1) **Species** are **resources** for lots of things that **humans need**, e.g. **rainforests** contain species that provide things like **drugs**, **clothes** and **food**. If the species and their habitats **aren't** conserved, the resources that we use now will be **lost**. Resources that **may be useful** in the **future** could also be **lost**.

2) Some people think we should conserve species simply because it's the **right thing to do**, e.g. most people think organisms have a **right to exist**, so they shouldn't become extinct as a result of **human activity**.

3) Many species and habitats bring **joy** to lots of people because they're **attractive** to **look at**. The species and habitats may be **lost** if they **aren't** conserved, so **future generations** won't be able to enjoy them.

4) Conserving species and habitats can help to prevent **climate change**. E.g. when trees are **burnt**, CO_2 is **released** into the atmosphere, which contributes to global warming. If they're conserved, this **doesn't happen**.

5) Conserving species and habitats helps to **prevent** the **disruption** of **food chains**. Disruption of food chains could mean the **loss** of **resources**. E.g. some species of **bear feed** on **salmon**, which feed on **herring** — if the number of herring **decreases** it can affect **both** the salmon and the bear populations.

Not everyone agrees with every conservation measure though — there's often **conflict** when conservation **affects people's livelihoods**, e.g. conservation of the Siberian tiger in Russia affects people who make money from killing the tigers and selling their fur (there's conflict between the conservationists and the hunters).

There are Many Different Ways to Conserve Species and Habitats

Different species and habitats need to be conserved in **different ways**.
Here are a few examples of **some** of the different **conservation methods** that can be used.

1) Plants can be Conserved using Seedbanks

1) A **seedbank** is a **store** of lots of **seeds** from lots of **different plant species**.

2) They help to conserve species by storing the seeds of **endangered** plants.

3) They also help to conserve **different varieties** of each species by storing a **range** of seeds from plants with **different characteristics**, e.g. seeds from tall sunflowers and seeds from short sunflowers.

4) If the plants become **extinct** in the wild the stored seeds can be used to **grow new plants**.

5) Seedbanks are a **good way** of conserving plant species — **large numbers** of species can be conserved because seeds don't need **much space**. Seeds can also be **stored anywhere** and for a **long time**, as long as it's **cool** and **dry**.

6) But there are **disadvantages** — the seeds have to be regularly tested to see if they're still **viable** (whether they can grow into a plant), which can be **expensive** and **time-consuming**.

The seedbank — 0% APR on branch transfers.

2) Fish species can be Conserved using Fishing Quotas

1) **Fishing quotas** are **limits** to the **amount** of certain fish species that fishermen are **allowed** to **catch**.

2) **Scientists** study different species and decide **how big** their populations need to be for them to **maintain** their numbers. Then they decide **how many** it's **safe** for fishermen to take without reducing the population **too much**.

3) **International agreements** are made (e.g. the Common Fisheries Policy in the EU) that state the **amount of** fish **each country** can take, and **where** they're allowed to take them from.

4) Fishing quotas help to **conserve** fish species by **reducing** the numbers that are **caught** and **killed**, so the populations aren't **reduced** too much and the species aren't at risk from becoming **extinct**.

5) There are **problems** with fishing quotas though — many fishermen **don't agree** with the scientists who say that the fish numbers are **low**. Some also think introducing quotas will cause **job losses**.

Conservation

3) Animals can be Conserved using Captive Breeding Programmes

1) Captive breeding programmes involve breeding animals in **controlled environments**.

2) Species that are **endangered**, or already **extinct in the wild**, can be **bred** in captivity to help **increase their numbers**, e.g. pandas are bred in captivity because their numbers are **critically low** in the wild.

3) There are some **problems** with captive breeding programmes though, e.g. animals can have **problems breeding** outside their **natural habitat**, which can be hard to **recreate** in a zoo. For example, pandas don't reproduce as **successfully** in captivity as they do in the wild.

4) Animals bred in captivity can be **reintroduced to the wild**. This **increases** their **numbers** in the wild, which can help to conserve their **numbers** or bring them **back** from the **brink of extinction**.

5) Reintroducing animals into the wild can cause **problems** though, e.g. reintroduced animals could bring **new diseases** to habitats, **harming** other species **living there**.

No way, I'm not breeding with him. He's ugly and his breath smells of bamboo.

4) Any organism can be Conserved by Relocation

1) **Relocating** a species means **moving** a population of a species to a **new location** because they're directly under **threat**, e.g. from poaching, or the **habitat** they're living in is under threat, e.g. from rising sea levels.

2) The species is moved to an area where it's **not at risk** (e.g. a protected national park, see below), but with a **similar environment** to where it's come from, so the species is still able to **survive**.

3) It's often used for species that only exist in **one place** (if that population **dies out**, the species will be **extinct**).

4) It helps to **conserve** species because they're relocated to a place where they're **more likely** to **survive**, so their numbers may **increase**.

5) Relocating species can cause **problems** though, e.g. native species in the new area may be **out-competed** by the species that's moved in and become **endangered** themselves.

5) Habitats can be Conserved using Protected Areas

1) **Protected areas** such as **national parks** and **nature reserves** protect habitats (and so protect the **species** in them) by **restricting urban development**, **industrial development** and **farming**.

2) Habitats in **protected areas** can be **managed** to conserve them, e.g. by **coppicing** — **cutting** down trees in a way that lets them **grow back**, so they don't need to be **replanted**. This helps to conserve the woodland, but allows some wood to be **harvested**.

3) There are **problems** with using protected areas to conserve habitats though, e.g. national parks are also used as **tourist destinations** (many are **funded** by **revenue** from the tourists that visit). This means there's conflict between the need to **conserve** the habitats and the need to allow people to **visit** and **use** them.

Practice Questions

Q1 Suggest why conservation of species and habitats is important for humans.

Q2 What is a seedbank?

Exam Question

Q1 A conservationist has argued that the deforestation of tropical rainforests will have terrible consequences for human beings and the environment.

a) Use your knowledge to outline the reasons for the conservation of tropical rainforests. [6 marks]

b) Name two suitable methods that could be used to conserve species from a tropical rainforest. [2 marks]

Captive breeding — you will procreate, or else...

There's lots of debate about conservation — what should be conserved and what's the best way to do it. That means there's a chance to get top marks in your exam — examiners love it when you talk about both sides of something. For example, even if you don't agree with captive breeding you need to say that it's got both positive and negative points.

Conservation Evidence and Data

And now my pretties, it's time for some data. Mwa ha ha...

You May Have to **Evaluate Evidence** and **Data** About **Conservation Issues**

You need to be able to **evaluate** any **evidence** or **data** about **conservation** projects and research that the examiners throw at you — so here's an example I made earlier:

In recent years, **native British bluebells** have become **less common** in woodland areas. It's thought that this is due to the presence of **non-native Spanish bluebells**, which compete with the native species for a **similar niche**. An experiment was carried out to see if **removing** the invasive Spanish species would help to **conserve** the native species. Each year for 15 years the **percentage cover** of native species was estimated in a **50 m by 50 m** area of **woodland** using random sampling and 250, **1 m² quadrats**. After five years, **all** the Spanish bluebells were **removed**. A **similar sized** control woodland in which the Spanish bluebells remained **untouched** was also studied. The results are shown on the right. You might be asked to:

1) **Describe the data:**

   - For the first **five years**, the **percentage cover** of **native bluebells fell** from **50%** to around **25%**. After the Spanish species was **removed**, it **increased** from around **25%** to around **45%** in **ten years**.
   - The **control experiment** shows a fairly **steady drop** in native bluebell percentage cover from **60%** to **20%** over the 15 years.

2) **Draw conclusions:**

   The removal of Spanish bluebells **resulted** in an **increase** in the percentage cover of **native bluebells** over a **ten year period**. This suggests that the **recent decrease** in native British bluebells is due to **competition** with the Spanish bluebells.

3) **Evaluate the method:**

   - The effects of some **other variables** (e.g. **changing weather**) were **removed** by the **control experiment**, where the percentage cover of native bluebells continued to fall throughout the 15 year study. This makes the data **more reliable**.
   - The **study area** and **sample size** were quite **large**, giving **more accurate** data.
   - **Random sampling** removed bias — the data's **more likely** to be an **accurate estimate** of the **whole area**.

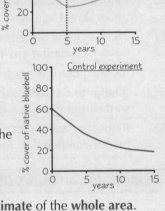

There's more about interpreting data on pages 106-108.

You Need to be Able to **Consider Conflicting Evidence**

1) The **evidence** from **one study** alone **wouldn't usually be enough** to conclude that there's a **link** between decreasing percentage cover of native bluebells, and the presence of Spanish bluebells.

2) **Similar studies** would be carried out to **investigate** the link. If these studies came to the **same conclusion**, the conclusion would become **increasingly accepted**.

3) Sometimes studies come up with **conflicting evidence** though — evidence that leads to a **different conclusion** than other studies. For example:

Another study was carried out to **investigate** the effect on native bluebells of **removing** Spanish bluebells. It was **similar** to the study above except a **20 m by 20 m** area was sampled using a random sample of **20 quadrats**, and **no control** woodland was used. You might be asked to:

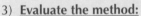

1) **Describe the data:**

   In the first five years, the **percentage cover** of **native bluebells fell** from **50%** to around **25%**. After the Spanish species was **removed**, it **kept decreasing** to around **15%** after the **full 15 years**.

2) **Draw conclusions:**

   The **removal** of the Spanish bluebells had **no effect** on the **decreasing** percentage cover of native bluebells — which **conflicts** with the study above.

3) **Evaluate the method:**

   - There **wasn't** a **control** woodland, so the **continuing decrease** in native bluebell cover after the removal of the Spanish bluebells could be due to **another factor**, e.g. cold weather in years 5-10.
   - The **study area** and **sample size** were quite small, giving a **less accurate** total percentage cover.

Conservation Evidence and Data

Conservation Relies on Science to Make Informed Decisions

1) Scientists carry out **research** to provide **information** about conservation issues.

2) This information can then be used to make **informed decisions** about **which** species and habitats **need** to be conserved, and the **best way** to conserve them.

3) For example, the study at the **top of the previous page** showed that **native bluebell** coverage increased after the removal of **Spanish bluebells**, which **suggests** that the decrease in native bluebell coverage is due to **competition** with the Spanish species. It provides evidence that there's a **conservation issue** (native bluebells are decreasing) and a way to **solve it** (remove the Spanish species).

Many conservation **decisions** have been made using the results of **scientific research** — take a look at these examples:

Scientific results	Decision
Between 1970 and 1989 the number of African elephants dropped from around 3 million to around 50 000 because they were being hunted for their ivory tusks.	In 1989, the Convention on International Trade in Endangered Species banned ivory trade to end the demand for elephant tusks, so that fewer elephants would be killed for their tusks.
The commonly used pesticide DDT was found to have contributed to the loss of half the peregrine falcon population in the UK in the 1950s and 1960s. DDT built up in the food chain and caused the falcon eggs to have thin shells. This meant the eggs were crushed and the chicks weren't hatched.	The use of DDT as a pesticide was banned in the UK in 1984 to try to conserve and increase peregrine falcon numbers.
The numbers of some species of sea turtle have dropped so low that they're now endangered. Many eggs are removed from the beaches by poachers before the turtles hatch and reach the sea.	Conservation agencies have set up hatching programmes where eggs are taken away from beaches and looked after until they hatch. The young turtles are then released into the sea.
A reduction in the size of hedgerows in farmers' fields was found to cause a decrease in biodiversity in the British countryside.	The government provides subsidies to encourage farmers to plant hedgerows and leave margins of ground unharvested around fields. This increases the size of hedgerows and conserves biodiversity.
Whale numbers were found to have dropped massively due to whale hunting.	Commercial whaling was banned in 1986 by the International Whaling Commission in order to conserve whale numbers.

Practice Questions

Q1 What is conflicting evidence?

Q2 Give one example of scientific evidence that has informed decision-making about conservation issues.

Q3 Give one example of a decision that has been made as a result of scientific evidence about conservation issues.

Exam Question

Q1 The graph shows the stock of spawning cod in the North Sea and the rate of mortality caused by fishing since 1960.

a) Describe the results shown by the graph. [4 marks]

b) Suggest a conclusion that could be drawn from the graph. [2 marks]

c) Scientists have stated that 150 000 tonnes is the minimum stock needed to preserve a cod population. In which year did cod stocks first fall below this level? [1 mark]

d) How might this data be used to make informed decisions about the conservation of cod stocks? [1 mark]

I'm considering conflict after these pages, I tell you...

Ah hah ha, aaaaah ha ha ha... oh, I think I need to stop my evil laugh now. I quite enjoyed that. Evaluating evidence and data's an important nut to crack — you might have to do it in your exam for conservation or for another topic altogether.

Inheritance

If you've ever wondered what causes colour blindness, how gender is controlled or how genetic diseases are passed on, then this is the section for you. If you've never wondered this and don't really care — tough. You still need to know it.

You Need to Know These Genetic Terms

'Codes for' means 'contains the instructions for'.

TERM	DESCRIPTION
Gene	A sequence of bases on a DNA molecule that codes for a protein (polypeptide), which results in a characteristic, e.g. the gene for eye colour.
Allele	A different version of a gene. Most plants and animals, including humans, have two alleles of each gene, one from each parent. The order of bases in each allele is slightly different — they code for different versions of the same characteristic. They're represented using letters, e.g. the allele for brown eyes (B) and the allele for blue eyes (b).
Genotype	The genetic constitution of an organism — the alleles an organism has, e.g. BB, Bb or bb for eye colour.
Phenotype	The expression of the genetic constitution and its interaction with the environment — an organism's characteristics, e.g. brown eyes.
Dominant	An allele whose characteristic appears in the phenotype even when there's only one copy. Dominant alleles are shown by a capital letter. E.g. the allele for brown eyes (B) is dominant — if a person's genotype is Bb or BB, they'll have brown eyes.
Recessive	An allele whose characteristic only appears in the phenotype if two copies are present. Recessive alleles are shown by a lower case letter. E.g. the allele for blue eyes (b) is recessive — if a person's genotype is bb, they'll have blue eyes.
Codominant	Alleles that are both expressed in the phenotype — neither one is recessive, e.g. the alleles for haemoglobin (see page 43).
Locus	The fixed position of a gene on a chromosome. Alleles of a gene are found at the same locus on each chromosome in a pair.
Homozygote	An organism that carries two copies of the same allele, e.g. BB or bb.
Heterozygote	An organism that carries two different alleles, e.g. Bb.

Genetic Diagrams Show the Possible Genotypes of Offspring

Individuals have **two alleles** for **each gene**. **Gametes** (sex cells) contain only **one allele** for each gene. When gametes from two parents fuse together, the alleles they contain form the **genotype** of the **offspring** produced. **Genetic diagrams** can be used to **predict** the **genotypes** and **phenotypes** of the offspring produced if two parents are **crossed** (bred). You need to know how to use genetic diagrams to predict the results of various crosses, including **monohybrid crosses**.

Monohybrid inheritance is the inheritance of a **single characteristic** (gene) controlled by **different alleles**. **Monohybrid crosses** show the **likelihood** of alleles (and so different versions of the characteristic) being **inherited** by offspring of particular parents. The genetic diagram below shows how **wing length** is inherited in fruit flies:

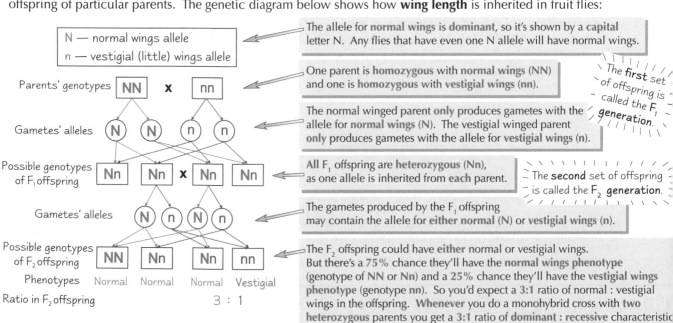

N — normal wings allele
n — vestigial (little) wings allele

Parents' genotypes: NN x nn

Gametes' alleles: N N n n

Possible genotypes of F₁ offspring: Nn Nn x Nn Nn

Gametes' alleles: N n N n

Possible genotypes of F₂ offspring: NN Nn Nn nn

Phenotypes: Normal Normal Normal Vestigial

Ratio in F₂ offspring: 3 : 1

The allele for **normal wings** is **dominant**, so it's shown by a **capital** letter N. Any flies that have even one N allele will have normal wings.

One parent is **homozygous** with **normal wings** (NN) and one is **homozygous** with **vestigial wings** (nn).

The normal winged parent **only** produces gametes with the allele for **normal wings** (N). The vestigial winged parent **only** produces gametes with the allele for **vestigial wings** (n).

*The **first** set of offspring is called the F₁ generation.*

All F₁ offspring are **heterozygous** (Nn), as one allele is inherited from **each** parent.

*The **second** set of offspring is called the F₂ generation.*

The gametes produced by the F₁ offspring may contain the allele for **either normal** (N) or **vestigial wings** (n).

The F₂ offspring could have **either** normal or vestigial wings. But there's a **75%** chance they'll have the **normal wings phenotype** (genotype of NN or Nn) and a **25%** chance they'll have the **vestigial wings phenotype** (genotype nn). So you'd expect a **3:1** ratio of normal : vestigial wings in the offspring. **Whenever** you do a monohybrid cross with **two heterozygous** parents you get a **3:1** ratio of **dominant : recessive** characteristic.

Inheritance

A **Punnett square** is just another way of showing a **genetic diagram** — they're also used to predict the **genotypes** and **phenotypes** of offspring. The Punnett squares below show the same crosses from the previous page:

1) First work out the alleles the **gametes** would have.

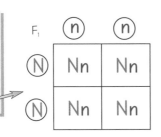

3) Then **cross the gametes' alleles of the F₁ generation** to show the possible **genotypes** of the F₂ generation. The Punnett square shows a **75%** chance that offspring will have **normal wings** and a **25%** chance that they'll have **vestigial wings**, i.e. a **3:1 ratio**.

2) Next **cross the parents' gametes** to show the possible genotypes of the F₁ generation — all heterozygous, Nn.

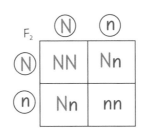

- 1 in 4 chance of offspring having the genotype NN (normal wings)
- 2 in 4 chance of offspring having the genotype Nn (normal wings)
- 1 in 4 chance of offspring having the genotype nn (vestigial wings)
- So, phenotype ratio normal:vestigial = 3:1

Some Genes Have Codominant Alleles

You need to be able to work out genetic diagrams for codominant alleles too.

Occasionally, alleles show **codominance** — **both alleles** are expressed in the **phenotype, neither one** is recessive. One example in humans is the allele for **sickle-cell anaemia**:

1) People who are **homozygous** for **normal haemoglobin** (H^NH^N) don't have the disease.

2) People who are **homozygous** for **sickle haemoglobin** (H^SH^S) have **sickle-cell anaemia** — all their **blood cells** are **sickle-shaped** (crescent-shaped).

3) People who are **heterozygous** (H^NH^S) have an **in-between** phenotype, called the **sickle-cell trait** — they have **some** normal haemoglobin and some sickle haemoglobin. The two alleles are **codominant** because they're **both expressed** in the **phenotype**.

4) The **genetic diagram** on the right shows the possible offspring from **crossing** two parents with **sickle-cell trait (heterozygous)**.

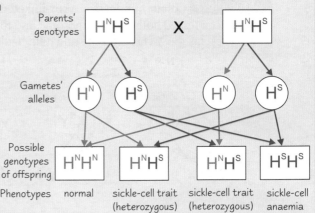

Practice Questions

Q1 What is meant by the term genotype?
Q2 What is meant by the term phenotype?
Q3 What is meant by the term codominance?

Exam Question

Q1 In pea plants, seed texture (round or wrinkled) is passed from parent to offspring by monohybrid inheritance. The allele for round seeds is represented by R and the allele for wrinkled seeds is represented by r.

a) Draw a genetic diagram to show the possible genotypes of F₁ offspring produced by crossing a homozygous round seed pea plant with a homozygous wrinkled seed pea plant. [3 marks]

b) What ratio of round to wrinkled seeds would you expect to see in the F₂ generation? [3 marks]

If there's a dominant revision allele I'm definitely homozygous recessive...

OK, so there are a lot of fancy words on these pages and yes, you do need to know them all. Sorry about that. But don't despair — once you've learnt what the words mean and know how genetic diagrams work it'll all just fall into place.

Inheritance

Now you know how these genetic diagram thingies work, you can use them to work out all kinds of clever stuff — even cleverer than the stuff you can already do. The crosses on these pages are a bit trickier, but nothing you can't handle.

Some **Characteristics** are **Sex-linked**

1) The genetic information for **gender** (**sex**) is carried on two **sex chromosomes**.

2) In mammals, **females** have **two X** chromosomes (XX) and **males** have **one X** chromosome and **one Y** chromosome (XY). The genetic diagram on the right shows how gender is **inherited**. The probability of having **male offspring** is **50%** and the probability of having **female offspring** is **50%**.

3) A **characteristic** is said to be **sex-linked** when the allele that codes for it is located on a **sex chromosome**.

4) The **Y** chromosome is **smaller** than the X chromosome and carries **fewer genes**. So most genes on the sex chromosomes are **only carried** on the X chromosome (called **X-linked** genes).

5) As **males** only have **one X chromosome** they often only have **one allele** for sex-linked genes. So because they **only** have one copy they **express** the **characteristic** of this allele even if it's **recessive**. This makes males **more likely** than females to show **recessive phenotypes** for genes that are sex-linked.

6) Genetic disorders caused by **faulty alleles** located on sex chromosomes include **colour blindness** and **haemophilia**. The faulty alleles for both of these disorders are carried on the X chromosome and so are called **X-linked disorders**. **Y-linked disorders** do exist but are **less common**.

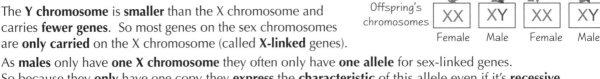

Example

Colour blindness is a **sex-linked disorder** caused by a faulty allele carried on the **X** chromosome. As it's sex-linked **both** the chromosome and the allele are **represented** in the **genetic diagram**, e.g. X^n, where X represents the **X chromosome** and n the **faulty allele** for **colour vision**. The **Y** chromosome doesn't have an allele for colour vision so is **just** represented by **Y**. **Females** would need **two copies** of the **recessive allele** to be colour blind, while **males** only need **one copy**. This means colour blindness is **much rarer** in **women** than **men**.

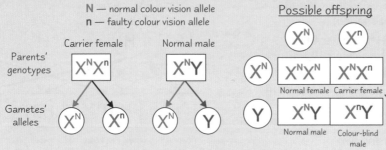

A carrier is a person carrying an allele which is not expressed in the phenotype but that can be passed on to offspring.

Some **Genes** Have **Multiple Alleles**

Inheritance is **more complicated** when there are **more than two** alleles of the same gene (**multiple alleles**).

Example

In the **ABO blood group system** in humans there are **three alleles** for blood type:

I^O is the allele for blood group **O**. I^A is the allele for blood group **A**. I^B is the allele for blood group **B**.

Allele I^O is **recessive**. Alleles I^A and I^B are **codominant** — people with genotype $I^A I^B$ will have blood group **AB**.

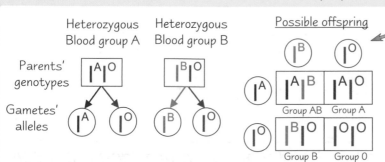

The genetic diagram shows a cross between a **heterozygous** person with blood group **A** and a **heterozygous** person with blood group **B**. Any offspring could have one of **four** different blood groups — **A**, **B**, **O** or **AB**.

Recessive blood groups are normally really rare, but it just so happens that loads of people in Britain are descended from people who were $I^O I^O$, so O's really common.

Inheritance

Genetic Pedigree Diagrams Show How Traits Run in Families

Genetic pedigree diagrams show an **inherited trait** (characteristic) in a group of **related individuals**. You might have to **interpret** genetic pedigree diagrams to work out the **genotypes** or **potential phenotypes** of individuals:

Example
Cystic fibrosis (CF) is an inherited disorder that's caused by a faulty **recessive** allele (f) — it codes for a **faulty chloride ion channel**. A person will only have the disorder if they're **homozygous** for the allele (ff) — they must inherit one recessive allele **from each parent**. If a person is **heterozygous** (Ff), they **won't** have CF but they'll be a **carrier**.

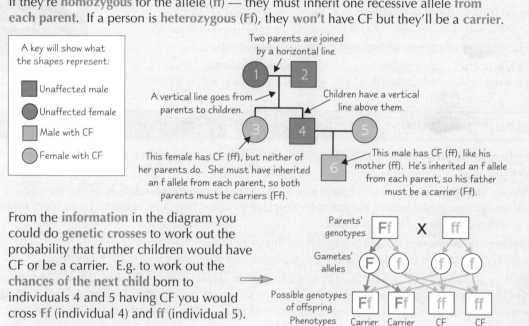

With a face this cute, Dillon knew he'd never have to worry about pedigree diagrams.

From the **information** in the diagram you could do **genetic crosses** to work out the probability that further children would have CF or be a carrier. E.g. to work out the **chances of the next child** born to individuals 4 and 5 having CF you would cross **Ff** (individual 4) and **ff** (individual 5).

Practice Questions

Q1 What is a sex-linked gene?

Q2 What is a carrier?

Q3 What do pedigree diagrams show?

Exam Questions

Q1 Haemophilia A is a sex-linked genetic disorder caused by a recessive allele carried on the X chromosome. Explain why haemophilia A is more common in males than females. [3 marks]

Q2 Using a genetic diagram, show the probability of a heterozygous person with blood group A and a homozygous person with blood group B having a child with blood group B. [4 marks]

Q3 ADA deficiency is an inherited metabolic disorder caused by a recessive allele (a). Use the genetic pedigree diagram above to answer the following questions:

a) Give the possible genotype(s) of individual 2. [1 mark]

b) What is the genotype of individual 6? Explain your answer. [2 marks]

c) What is the probability that the next child born to individuals 5 and 6 will have ADA deficiency? Show your working. [4 marks]

Sex-linkage and multiple alleles — it's all starting to sound a little bit kinky...

So sex-linked characteristics are characteristics linked to your sex, more than two alleles for one gene are called multiple alleles, and genetic pedigree diagrams are just family trees with a few genotypes thrown in. Another two pages ticked off.

The Hardy-Weinberg Principle

Sometimes you need to look at the genetics of a whole population, rather than a cross between just two individuals. And that's where those spiffing fellows Hardy and Weinberg come in...

Members of a Population Share a Gene Pool

1) A **species** is defined as a group of **similar organisms** that can **reproduce** to give **fertile offspring**.

2) A **population** is a group of organisms of the **same species** living in a **particular area**.

3) Species can exist as **one** or **more populations**, e.g. there are populations of the American black bear (*Ursus americanus*) in parts of America and in parts of Canada.

4) The **gene pool** is the complete range of **alleles** present in a **population**.

5) How **often** an **allele occurs** in a population is called the **allele frequency**. It's usually given as a **percentage** of the total population, e.g. 35%, or a **number**, e.g. 0.35.

Yogi wanted everyone to know what population he was in.

The Hardy-Weinberg Principle Predicts That Allele Frequencies Won't Change

1) The **Hardy-Weinberg principle** predicts that the **frequencies** of **alleles** in a population **won't change** from **one generation** to the **next**.

2) But this prediction is **only true** under **certain conditions** — it has to be a **large population** where there's **no immigration**, **emigration**, **mutations** or **natural selection** (see p. 48). There also needs to be **random mating** — all possible genotypes can breed with all others.

3) The **Hardy-Weinberg equations** (see below) are based on this principle. They can be used to **estimate the frequency** of particular **alleles** and **genotypes** within populations.

4) The Hardy-Weinberg equations can also be used to test whether or not the Hardy-Weinberg principle **applies** to **particular alleles** in **particular populations**, i.e. to test whether **selection** or any **other factors** are **influencing** allele frequencies — if frequencies **do change** between generations in a large population then there's a pressure of some kind (see next page).

The Hardy-Weinberg Equations Can be Used to...

...Predict Allele Frequency...

1) You can **figure out** the frequency of one allele if you **know the frequency of the other**, using this equation:

$$p + q = 1$$

Where: **p** = the **frequency** of the **dominant** allele
q = the **frequency** of the **recessive** allele

The **total frequency** of **all possible alleles** for a characteristic in a certain population is **1.0**. So the frequencies of the **individual alleles** (the dominant one and the recessive one) must **add up to 1.0**.

2) E.g. a species of plant has either **red** or **white** flowers. Allele **R** (red) is **dominant** and allele **r** (white) is **recessive**. If the frequency of **R** is **0.4**, then the frequency of **r** is 1 − 0.4 = **0.6**.

...Predict Genotype Frequency...

Make sure you learn both Hardy-Weinberg equations.

1) You can **figure out** the frequency of one genotype if you **know the frequencies of the others**, using this equation:

$$p^2 + 2pq + q^2 = 1$$

Where p^2 = the **frequency** of the **homozygous dominant genotype**
$2pq$ = the **frequency** of the **heterozygous genotype**
q^2 = the **frequency** of the **homozygous recessive genotype**

The **total frequency** of **all possible genotypes** for one characteristic in a certain population is **1.0**. So the frequencies of the **individual genotypes** must **add up to 1.0**.

2) E.g. If there are **two alleles** for **flower colour** (R and r), there are **three possible genotypes** — **RR**, **Rr** and **rr**. If the frequency of genotype **RR** (p^2) is **0.34** and the frequency of genotype **Rr** ($2pq$) is **0.27**, the frequency of genotype **rr** (q^2) must be 1 − 0.34 − 0.27 = **0.39**.

The Hardy-Weinberg Principle

...*Predict* the *Percentage* of a *Population* that has a *Certain Genotype*...

Example

The **frequency** of **cystic fibrosis** (genotype **ff**) in the UK is currently approximately **1 birth in 2000**.
From this information you can estimate the **proportion** of people in the UK that are cystic fibrosis **carriers (Ff)**.
To do this you need to find the **frequency of heterozygous genotype Ff**, i.e. **2pq**, using **both** equations:

First calculate q:
- Frequency of cystic fibrosis (homozygous recessive, ff) is 1 in 2000
- $ff = q^2 = 1 \div 2000 = 0.0005$
- So, $q = \sqrt{0.0005} = 0.022$

Next calculate p:
- using $p + q = 1$, $p = 1 - q$
- $p = 1 - 0.022$
- $p = 0.978$

Then calculate 2pq:
- $2pq = 2 \times 0.978 \times 0.022$
- $2pq = 0.043$

The **frequency** of genotype Ff is **0.043**, so the **percentage** of the UK population that are **carriers** is **4.3%**.

...and *Show if External Factors* are *Affecting Allele Frequency*

Example

If the **frequency** of **cystic fibrosis** is measured **50 years later** it might be found to be **1 birth in 3000**.
From this information you can estimate the **frequency** of the **recessive allele** (f) in the population, i.e. **q**.

To calculate q:
- Frequency of cystic fibrosis (homozygous recessive, ff) is 1 in 3000
- $ff = q^2 = 1 \div 3000 = 0.00033$
- So, $q = \sqrt{0.00033} = 0.018$

The frequency of the recessive allele is now **0.018**, compared to **0.022** currently (see above). As the frequency of the allele has **changed** between generations the **Hardy-Weinberg principle doesn't apply** so there must have been some **factors** affecting **allele frequency**, e.g. **immigration, migration, mutations** or **natural selection**.

Practice Questions

Q1 What is a population?

Q2 What is a gene pool?

Q3 What conditions are needed for the Hardy-Weinberg principle to apply?

Q4 Which term represents the frequency of the homozygous recessive genotype in the Hardy-Weinberg equations?

Q5 Which term represents the frequency of the heterozygous genotype in the Hardy-Weinberg equations?

Exam Question

Q1 Cleft chins are controlled by a single gene with two alleles. The allele coding for a cleft chin (C) is dominant over the allele coding for a non-cleft chin (c). In a particular population the frequency of the homozygous dominant genotype for cleft chin is 0.14.

a) What is the frequency of the recessive allele in the population? [3 marks]

b) What is the frequency of the homozygous recessive genotype in the population? [1 mark]

c) What percentage of the population have a cleft chin? [2 marks]

This stuff's surely not that bad — Hardly worth Weining about...

Two equations that you absolutely have to know — so learn 'em. And whilst you're at it make sure that you learn what each of the terms means as well. You'll feel like a right wally if you know that $p^2 + 2pq + q^2 = 1$ but haven't got a clue what p^2, 2pq and q^2 stand for. It's the kind of stuff that falls out of your head really easily so learn it, learn it, learn it.

Allele Frequency and Speciation

I'm sure my brother's from a totally different species, boom boom — it's an oldie but a goodie.

Allele Frequency is Affected by Differential Reproductive Success

1) Sometimes the **frequency** of an **allele** within a population **changes**. This can happen when the allele codes for a characteristic that **affects** the **chances** of an organism **surviving**.

2) Not all individuals are as likely to **reproduce** as each other. There's **differential reproductive success** in a population — individuals that have an allele that **increases** their **chance of survival** are **more likely** to **survive**, **reproduce** and **pass on** their genes (including the **beneficial** allele), than individuals with different alleles.

3) This means that a **greater proportion** of the next generation **inherit** the **beneficial allele**.

4) They, in turn, are **more likely** to **survive**, **reproduce** and **pass on** their genes.

5) So the **frequency** of the beneficial allele **increases** from generation to generation.

6) This process is called **natural selection**.

Different Types of Natural Selection Lead to Different Frequency Patterns

Stabilising selection and **directional selection** are **types** of **natural selection** that affect **allele frequency** in different ways:

Stabilising selection is where individuals with alleles for characteristics towards the **middle** of the range are more likely to **survive** and **reproduce**. It occurs when the environment **isn't changing**, and it **reduces the range** of possible **phenotypes**.

EXAMPLE In any **mammal population** there's a **range** of **fur length**. In a **stable climate**, having fur at the **extremes** of this range **reduces** the **chances** of **surviving** as it's harder to maintain the **right body temperature**. Animals with alleles for **average fur length** are the **most** likely to **survive**, **reproduce** and **pass on** their alleles. So these alleles **increase** in **frequency**. The **proportion** of the **population** with **average fur length increases** and the **range** of fur lengths **decreases**.

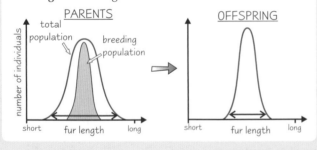

Directional selection is where individuals with alleles for characteristics of an **extreme type** are more likely to **survive** and **reproduce**. This could be in response to an **environmental change**.

EXAMPLE Cheetahs are the **fastest** animals on land. It's likely that this characteristic was developed through **directional selection**, as individuals that have **alleles** for **speed** are **more likely** to **catch prey** than slower individuals. So they're **more likely** to **survive**, **reproduce** and **pass on** their alleles. Over time the **frequency** of alleles for **high speed increases** and the population becomes **faster**.

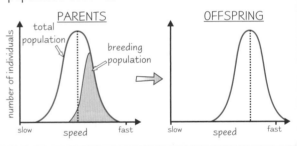

You Need to be able to Interpret Data Relating to the Effect of Selection

1) For example, there are **two common forms** of the peppered moth in the UK — one **dark coloured** and one **pale**.

2) The allele for **dark colouring** is **dominant** (M) over the allele for **pale colouring** (m).

3) The table shows how **allele frequency** in a population of **peppered moths** changed between 1852 and 1860 as the number of **coal-powered factories** in the area increased.

4) The frequency of the **m** allele **decreases** from **0.75** to **0.39** as the number of factories increases. So the frequency of the **M allele** must **increase** from **0.25** to **0.61** (remember: p + q = 1, see page 46).

5) As the allele frequencies are **changing**, it's likely there's selective pressure **for** dark colouring. This could be because of **pollution** — more factories means more pollution, which **darkens** buildings etc. The **dark coloured moths** would be better **camouflaged**, making them **more likely** to **survive**, **reproduce** and pass on **M**.

Year	Number of coal-powered factories	Frequency of m allele
1852	1	0.75
1854	3	0.70
1856	5	0.63
1858	7	0.47
1860	10	0.39

Allele Frequency and Speciation

Geographical Isolation and Natural Selection Lead to Speciation

1) **Speciation** is the development of a **new species**.

2) Speciation occurs when populations of the same species become **reproductively isolated**.

3) This can happen when a **physical barrier**, e.g. a flood or an earthquake, **divides** a population of a species, causing some individuals to become **separated** from the main population. This is known as **geographical isolation**.

4) Populations that are geographically separated will experience slightly **different conditions**. For example, there might be a **different climate** on each side of the physical barrier.

5) The populations will experience **different selective pressures** and so **different changes** in allele frequencies:

   - Different **alleles** will be **more advantageous** in the different populations. For example, if geographical separation places one population in a **colder climate** than before, **longer fur length** will be **beneficial**. **Directional selection** will then act on the **alleles** for fur length in this population, changing the frequency of the allele for **longer fur length**.

   - Allele frequencies will also change as **mutations** (see p. 84) will occur **independently** in each population.

6) The changes in allele frequency will lead to **differences** accumulating in the **gene pools** of the separated populations, causing changes in **phenotype frequencies**.

7) Eventually, individuals from the different populations will have changed so much that they won't be able to breed with one another to produce **fertile** offspring — they'll have become **reproductively isolated**.

Sandra's hair had caused her reproductive isolation.

8) The two groups will have become **separate species**.

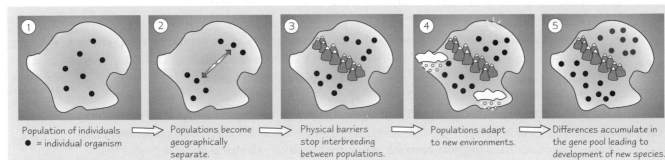

Population of individuals
● = individual organism
⟹ Populations become geographically separate.
⟹ Physical barriers stop interbreeding between populations.
⟹ Populations adapt to new environments.
⟹ Differences accumulate in the gene pool leading to development of new species.

Practice Questions

Q1 What is stabilising selection?

Q2 What is directional selection?

Q3 What is speciation?

Average Temp / °C	Frequency of h allele
22	0.11
21	0.13
19	0.19
18	0.20
16	0.23

Exam Question

Q1 The table above shows the results of an investigation into hair length in golden hamsters in a climate where the temperature is decreasing. Hair length is controlled by a single gene with two alleles. H represents the allele for short hair, which is dominant over the allele for long hair, represented by h.

a) Describe the relationship between the frequency of the recessive long hair allele and temperature. Suggest an explanation for this relationship. [4 marks]

b) What type of selection is responsible for this change in allele frequency? [1 mark]

Differential reproductive success — not PC, but it sorts the hot from the not...

And that's the end of the section. Whoopdeedoo. All that's left for you to do is learn these two pages — no small feat I grant you, as they cover a lot of stuff. Maybe have another quick flick through the section before you sign off from genetic variation entirely. You know it makes sense, and who knows — you might enjoy it... OK maybe not, but do it anyway.

Nervous and Hormonal Communication

Right, it's time to get your brain cells fired up and get your teeth stuck into a mammoth — a mammoth section, that is...

Responding to their Environment Helps Organisms Survive

1) **Animals increase** their **chances** of **survival** by **responding** to **changes** in their **external environment**, e.g. by **avoiding harmful environments** such as places that are too hot or too cold.

2) They also **respond** to **changes** in their **internal environment** to make sure that the **conditions** are always **optimal** for their **metabolism** (all the chemical reactions that go on inside them).

3) **Plants** also **increase** their **chances** of **survival** by **responding** to **changes** in their **environment** (see p. 66).

4) Any **change** in the internal or external **environment** is called a **stimulus**.

Receptors Detect Stimuli and Effectors Produce a Response

1) **Receptors detect stimuli** — they can be **cells** or **proteins** on **cell surface membranes**. There are **loads** of **different types** of receptors that detect **different stimuli**.

2) **Effectors** are cells that bring about a **response** to a **stimulus**, to produce an **effect**. Effectors include **muscle cells** and cells found in **glands**, e.g. the **pancreas**.

3) Receptors **communicate** with effectors via the **nervous system** or the **hormonal system**, or sometimes using **both**.

The Nervous System Sends Information as Electrical Impulses

1) The **nervous system** is made up of a **complex network** of cells called **neurones**. There are **three main types** of neurone:

- **Sensory neurones** transmit electrical impulses from **receptors** to the **central nervous system** (**CNS**).
- **Motor neurones** transmit electrical impulses from the **CNS** to **effectors**.
- **Relay neurones** transmit electrical impulses **between** sensory neurones and motor neurones.

Electrical impulses are also called nerve impulses.

2) A stimulus is detected by **receptor cells** and an **electrical impulse** is sent along a **sensory neurone**.

3) When an **electrical impulse** reaches the end of a neurone chemicals called **neurotransmitters** take the information across to the **next neurone**, which then sends an **electrical impulse** (see p. 57).

4) The **CNS processes** the information, **decides what to do** about it and sends impulses along **motor neurones** to an **effector**.

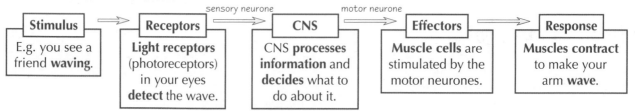

Stimulus		sensory neurone →	**CNS**	motor neurone →	**Effectors**		**Response**
E.g. you see a friend **waving**.		**Receptors** **Light receptors** (photoreceptors) in your eyes **detect** the wave.	CNS **processes** **information** and **decides** what to do about it.		Muscle cells are stimulated by the motor neurones.		**Muscles contract** to make your arm **wave**.

5) The **nervous system** is split into two different systems:

You don't need to learn the structure of the nervous system, but understanding it'll help with the rest of the section.

The **central nervous system** (**CNS**) — made up of the **brain** and the **spinal cord**.

The **peripheral nervous system** — made up of the neurones that connect the CNS to the **rest** of the **body**. It also has two different systems:

The **somatic nervous system** controls **conscious** activities, e.g. running and playing video games.

The **autonomic nervous system** controls **unconscious** activities, e.g. digestion. It's got two divisions that have **opposite effects** on the body:

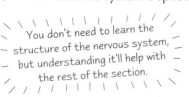

Harold thought it was about time his sympathetic nervous system took over.

The **sympathetic** nervous system gets the body **ready for action**. It's the '**flight or fight**' system.

The **parasympathetic** nervous system **calms** the body down. It's the '**rest and digest**' system.

Nervous and Hormonal Communication

Nervous System Communication is Localised, Short-lived and Rapid

1) When an **electrical impulse** reaches the end of a neurone, **neurotransmitters** are **secreted directly** onto **cells** (e.g. muscle cells) — so the nervous response is **localised**.

2) **Neurotransmitters** are **quickly removed** once they've done their job, so the response is **short-lived**.

3) Electrical impulses are **really fast**, so the response is **rapid** — this allows animals to **react quickly** to stimuli.

The Hormonal System Sends Information as Chemical Signals

1) The **hormonal system** is made up of **glands** and **hormones**:
   - A **gland** is a group of cells that are specialised to **secrete** a useful substance, such as a **hormone**. E.g. the **pancreas** secretes **insulin**.
   - **Hormones** are 'chemical messengers'. Many hormones are **proteins** or **peptides**, e.g. **insulin**. Some hormones are **steroids**, e.g. **progesterone**.

2) **Hormones** are **secreted** when a **gland** is **stimulated**:
   - Glands can be **stimulated** by a **change** in **concentration** of a specific **substance** (sometimes **another hormone**).
   - They can also be **stimulated** by **electrical impulses**.

3) Hormones **diffuse directly into** the **blood**, then they're **taken** around the body by the **circulatory system**.

4) They **diffuse out** of the blood **all over** the **body** but each hormone will only **bind** to **specific receptors** for that hormone, found on the membranes of some cells (called **target cells**).

5) The hormones trigger a **response** in the **target cells** (the **effectors**).

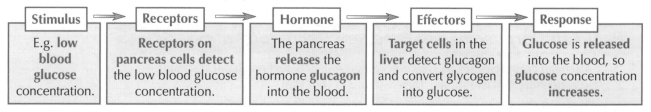

Stimulus	Receptors	Hormone	Effectors	Response
E.g. **low blood glucose** concentration.	**Receptors on pancreas cells** detect the low blood glucose concentration.	The pancreas **releases** the hormone **glucagon** into the blood.	**Target cells** in the **liver** detect glucagon and convert glycogen into glucose.	**Glucose is released** into the blood, so **glucose** concentration **increases**.

Hormonal System Communication is Slower, Long-lasting and Widespread

1) Hormones **aren't** released directly onto their target cells — they must **travel** in the **blood** to get there. This means that chemical communication (by hormones) is **slower** than electrical communication (by nerves).

2) They **aren't broken down as quickly** as neurotransmitters, so the **effects** of hormones can **last** for much **longer**.

3) Hormones are transported **all over** the **body**, so the response may be **widespread** if the target cells are widespread.

Practice Questions

Q1 Why do organisms respond to changes in their environment?

Q2 Give two types of effector.

Q3 How do hormones reach their target cells?

Exam Question

Q1 Bright light causes circular iris muscles in an animal's eyes to contract, which constricts the pupils and protects the eyes.

   a) Suggest why this response uses nervous communication rather than hormonal communication. [1 mark]

   b) Describe and explain the roles of receptors and effectors for this response. [5 marks]

Vacancy — talented gag writer required for boring biology topics...

Actually, this stuff is really quite fascinating once you realise just how much your body can do without you even knowing. Just sit back and relax, let your nerves and hormones do the work... Ah, apart from the whole revision thing — your body can't do that without you knowing, unfortunately. Get your head around these pages before you tackle the rest of the section.

Receptors

So now you know why organisms respond it's time for the (slightly less interesting but equally important) details... first up — receptors.

Receptors are Specific to One Kind of Stimulus

1) Receptors are **specific** — they only **detect one particular stimulus**, e.g. light, pressure or glucose concentration.

2) There are **many different types** of receptor that each detect a **different type of stimulus**.

3) Some receptors are **cells**, e.g. photoreceptors are receptor cells that connect to the nervous system. Some receptors are **proteins** on **cell surface membranes**, e.g. glucose receptors are proteins found in the cell membranes of some pancreatic cells.

4) Here's a bit more about how receptor cells that communicate information via the **nervous system** work:

   - When a nervous system receptor is in its **resting state** (not being stimulated), there's a **difference in voltage** (**charge**) between the **inside** and the **outside** of the cell — this is generated by ion pumps and ion channels (see p. 54). The **difference in voltage** across the membrane is called the **potential difference**.

   - The **potential difference** when a cell is at **rest** is called its **resting potential**. When a stimulus is detected, the cell membrane is **excited** and becomes **more permeable**, allowing **more ions** to move **in** and **out** of the cell — **altering** the **potential difference**. The **change** in **potential difference** due to a stimulus is called the **generator potential**.

   - A **bigger stimulus** excites the membrane more, causing a **bigger movement** of ions and a **bigger change** in potential difference — so a **bigger generator potential** is produced.

   - If the **generator potential** is **big enough** it'll trigger an **action potential** — an electrical impulse along a neurone (see p. 54). An action potential is only triggered if the generator potential reaches a certain level called the **threshold** level. Action potentials are all one size, so the **strength** of the **stimulus** is measured by the **frequency** of **action potentials**.

   - If the stimulus is **too weak** the generator potential **won't reach** the **threshold**, so there's **no action potential**.

Pacinian Corpuscles are Pressure Receptors in Your Skin

1) **Pacinian corpuscles** are **mechanoreceptors** — they detect **mechanical stimuli**, e.g. **pressure** and **vibrations**. They're found in your **skin**.

2) Pacinian corpuscles contain the end of a **sensory neurone**, imaginatively called a **sensory nerve ending**.

3) The sensory nerve ending is **wrapped** in loads of layers of connective tissue called **lamellae**.

4) When a Pacinian corpuscle is **stimulated**, e.g. by a tap on the arm, the lamellae are **deformed** and **press** on the **sensory nerve ending**.

5) This causes **deformation** of **stretch-mediated sodium channels** in the sensory neurone's cell membrane. The sodium ion channels **open** and **sodium ions diffuse into** the cell, creating a **generator potential**.

6) If the **generator potential** reaches the **threshold**, it triggers an **action potential**.

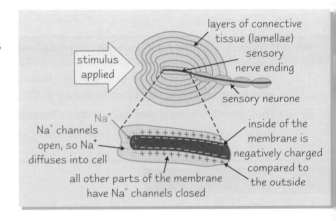

layers of connective tissue (lamellae)
sensory nerve ending
stimulus applied
sensory neurone
Na^+
Na^+ channels open, so Na^+ diffuses into cell
all other parts of the membrane have Na^+ channels closed
inside of the membrane is negatively charged compared to the outside

Photoreceptors are Light Receptors in Your Eye

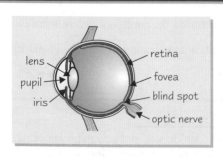

lens
pupil
iris
retina
fovea
blind spot
optic nerve

1) **Light** enters the eye through the **pupil**. The **amount** of light that enters is **controlled** by the muscles of the **iris**.

2) Light rays are **focused** by the **lens** onto the **retina**, which lines the inside of the eye. The retina contains **photoreceptor cells** — these **detect light**.

3) The **fovea** is an area of the retina where there are **lots** of **photoreceptors**.

4) **Nerve impulses** from the photoreceptor cells are carried from the **retina** to the **brain** by the **optic nerve**, which is a bundle of **neurones**. Where the optic nerve leaves the eye is called the **blind spot** — there **aren't** any **photoreceptor cells**, so it's **not sensitive** to **light**.

Receptors

Photoreceptors Convert Light into an Electrical Impulse

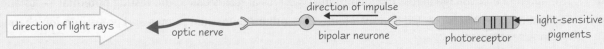

Light goes straight through the neurones to the photoreceptors.

1) **Light** enters the eye, hits the **photoreceptors** and is **absorbed** by **light-sensitive pigments**.

2) Light bleaches the pigments, causing a **chemical change** and altering the **membrane permeability** to **sodium**.

3) A **generator potential** is created and if it reaches the threshold, a nerve impulse is sent along a **bipolar neurone**.

4) Bipolar neurones connect **photoreceptors** to the **optic nerve**, which takes impulses to the **brain**.

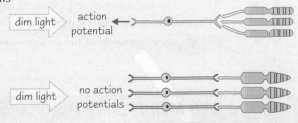

5) The human eye has **two types** of photoreceptor — **rods** and **cones**.

6) Rods are mainly found in the **peripheral** parts of the **retina**, and cones are found **packed together** in the **fovea**.

7) Rods only give information in **black and white** (monochromatic vision), but cones give information in **colour** (trichromatic vision). There are three types of cones — **red-sensitive**, **green-sensitive** and **blue-sensitive**. They're stimulated in **different proportions** so you see different colours.

Rods are More Sensitive, but Cones let you See More Detail

Sensitivity

- Rods are **very sensitive to light** (they fire action potentials in **dim light**). This is because **many rods** join **one neurone**, so many weak **generator potentials combine** to **reach** the **threshold** and trigger an action potential.

- Cones are **less sensitive** than rods (they only fire action potentials in **bright light**). This is because **one cone** joins **one neurone**, so it takes more light to reach the threshold and trigger an action potential.

Visual acuity (the ability to tell apart points that are close together)

- Rods give **low visual acuity** because **many rods** join the **same neurone**, which means light from two objects close together **can't** be told apart.

- Cones give **high visual acuity** because cones are **close together** and **one cone** joins **one neurone**. When light from two points hits two cones, two action potentials (one from **each cone**) go to the brain — so you can distinguish two points that are close together as **two separate points**.

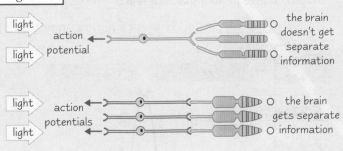

Practice Questions

Q1   What is a generator potential?

Q2   What is the threshold level of a receptor?

Exam Questions

Q1   Explain how a tap on the arm is converted into a nerve impulse.   [6 marks]

Q2   Explain how the human eye has both high sensitivity and high acuity.   [5 marks]

Pacinian corpuscles love deadlines — they work best under pressure...

Wow, loads of stuff here, so cone-gratulations if you manage to remember it all. Receptors are really important because without them you wouldn't be able to see this book, and without this book revision would be way trickier.

Nervous System — Neurones

Ah, on to the good stuff. Revision notepad at the ready, motor neurones fired up, OK — lights, camera, action potentials...

Neurone **Cell Membranes** are **Polarised** at **Rest**

1) In a neurone's **resting state** (when it's not being stimulated), the **outside** of the membrane is **positively charged** compared to the **inside**. This is because there are **more positive ions outside** the cell than inside.

2) So the membrane is **polarised** — there's a **difference in voltage** across it.

3) The difference in voltage across the membrane when it's at rest is called the **resting potential** — it's about **–70 mV**.

4) The resting potential is created and maintained by **sodium-potassium pumps** and **potassium ion channels** in a neurone's membrane:

Sodium-potassium pump

These pumps use **active transport** to move **three sodium ions** (Na$^+$) **out** of the neurone for every **two potassium ions** (K$^+$) moved **in**. ATP is needed to do this.

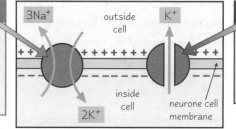

Potassium ion channel

These channels allow **facilitated diffusion** of **potassium ions** (K$^+$) **out** of the neurone, down their **concentration gradient**.

- The sodium-potassium pumps move **sodium ions out** of the neurone, but the membrane **isn't permeable** to **sodium ions**, so they **can't diffuse back in**. This creates a **sodium ion electrochemical gradient** (a **concentration gradient** of **ions**) because there are **more** positive sodium ions **outside** the cell than inside.

- The sodium-potassium pumps also move **potassium ions in** to the neurone, but the membrane **is permeable** to **potassium ions** so they **diffuse back out** through potassium ion channels.

- This makes the **outside** of the cell **positively charged** compared to the inside.

Neurone **Cell Membranes** Become **Depolarised** when they're **Stimulated**

A **stimulus** triggers other ion channels, called **sodium ion channels**, to **open**. If the stimulus is big enough, it'll trigger a **rapid change in potential difference**. The sequence of events that happen are known as an **action potential**:

Change in potential difference during an action potential

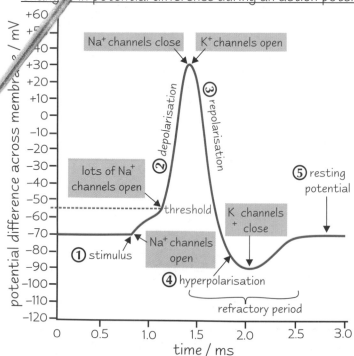

① **Stimulus** — this **excites** the neurone cell membrane, causing **sodium ion channels** to **open**. The membrane becomes **more permeable** to sodium, so **sodium ions diffuse into** the neurone down the sodium ion electrochemical gradient. This makes the **inside** of the neurone **less negative**.

② **Depolarisation** — if the potential difference reaches the **threshold** (around **–55 mV**), **more sodium ion channels open**. More sodium ions **diffuse into** the neurone.

Nervous System — Neurones

③ **Repolarisation** — at a potential difference of around **+30 mV** the **sodium ion channels close** and **potassium ion channels open**. The membrane is **more permeable** to potassium so **potassium ions diffuse out** of the neurone down the potassium ion concentration gradient. This starts to get the membrane **back** to its **resting potential**.

④ **Hyperpolarisation** — **potassium ion channels** are **slow to close** so there's a slight **'overshoot'** where too many potassium ions diffuse out of the neurone. The potential difference becomes **more negative** than the **resting potential** (i.e. less than –70 mV).

⑤ **Resting potential** — the ion channels are **reset**. The **sodium-potassium pump** returns the membrane to its **resting potential** and maintains it until the membrane's excited by another stimulus.

After an **action potential**, the neurone cell membrane **can't** be **excited** again straight away. This is because the ion channels are **recovering** and they **can't** be made to **open** — sodium ion channels are **closed** during repolarisation and **potassium ion channels** are **closed** during hyperpolarisation. This period of recovery is called the **refractory period**.

The **Action Potential** Moves **Along** the **Neurone**

1) When an **action potential** happens, some of the **sodium ions** that enter the neurone **diffuse sideways**.

2) This causes **sodium ion channels** in the next region of the neurone to **open** and **sodium ions diffuse into** that part.

3) This causes a **wave of depolarisation** to travel along the neurone.

4) The **wave** moves **away** from the parts of the membrane in the **refractory period** because these parts **can't fire** an action potential.

It's like a Mexican wave travelling through a crowd — sodium ions rushing inwards causes a wave of activity along the membrane.

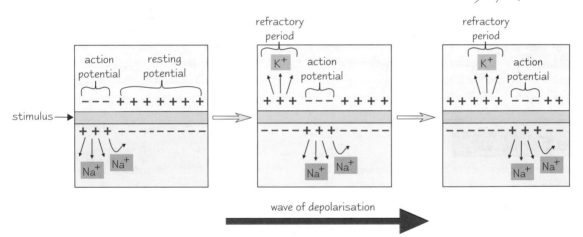

wave of depolarisation

The **Refractory Period** Produces **Discrete Impulses**

1) During the **refractory period**, **ion channels** are **recovering** and **can't** be **opened**.

2) So the refractory period acts as a **time delay** between one action potential and the next. This makes sure that **action potentials don't overlap** but pass along as **discrete** (separate) **impulses**.

3) The refractory period also makes sure **action potentials** are **unidirectional** (they only travel in **one direction**).

Action Potentials have an All-or-Nothing Nature

1) Once the threshold is reached, an action potential will **always fire** with the **same change in voltage**, no matter how big the stimulus is.

2) If **threshold isn't reached**, an action potential **won't fire** — this is the **all-or-nothing** nature of action potentials.

3) A **bigger stimulus** won't cause a bigger action potential but it will cause them to fire **more frequently**.

Nervous System — Neurones

Three Factors Affect the Speed of Conduction of Action Potentials

① Myelination

1) Some neurones are **myelinated** — they have a **myelin sheath**.

2) The myelin sheath is an **electrical insulator**.

3) It's made of a type of cell called a **Schwann cell**.

4) Between the Schwann cells are tiny patches of **bare membrane** called the **nodes of Ranvier**. **Sodium ion channels** are **concentrated** at the nodes.

5) In a **myelinated** neurone, **depolarisation** only happens at the **nodes of Ranvier** (where sodium ions can get through the membrane).

6) The neurone's **cytoplasm conducts** enough electrical charge to **depolarise** the **next node**, so the impulse '**jumps**' from node to node.

7) This is called **saltatory conduction** and it's **really fast**.

8) In a **non-myelinated** neurone, the impulse travels as a **wave** along the **whole length** of the axon membrane.

9) This is **slower** than saltatory conduction (although it's still pretty quick).

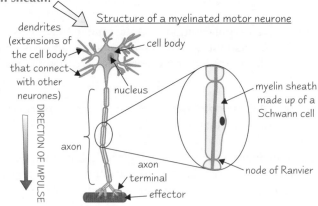

Structure of a myelinated motor neurone

dendrites (extensions of the cell body that connect with other neurones)

cell body

nucleus

DIRECTION OF IMPULSE

axon

axon terminal

effector

myelin sheath made up of a Schwann cell

node of Ranvier

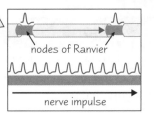

nodes of Ranvier

nerve impulse

You need to learn the structure of a myelinated motor neurone.

② Axon diameter

Action potentials are conducted **quicker** along axons with **bigger diameters** because there's **less resistance** to the **flow of ions** than in the cytoplasm of a smaller axon. With less resistance, **depolarisation reaches** other parts of the neurone cell membrane **quicker**.

③ Temperature

The speed of conduction increases as the **temperature increases** too, because **ions diffuse faster**. The speed only increases up to around **40 °C** though — after that the **proteins** begin to **denature** and the speed decreases.

Practice Questions

Q1 Give one function of the refractory period.

Q2 What is meant by the 'all-or-nothing' nature of action potentials?

Q3 What is the function of Schwann cells on a neurone?

Q4 Name three features of axons that speed up the conduction of action potentials.

These questions cover pages 54-56.

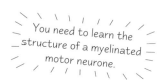

Exam Questions

Q1 The graph on the right shows an action potential across an axon membrane following the application of a stimulus.

a) What label should be added at point A? [1 mark]

b) Explain what causes the change in potential difference between point A and point B. [3 marks]

c) A stimulus was applied at 1.5 ms, but failed to produce an action potential. Suggest why. [2 marks]

Q2 Multiple sclerosis is a disease of the nervous system characterised by damage to the myelin sheaths of neurones. Explain how this will affect the transmission of action potentials. [5 marks]

I'm feeling a bit depolarised after all that...

All this stuff about neurones can be a bit tricky to get your head around at first. Take your time and try scribbling it all down a few times till it starts to make some kind of sense. Basically, neurones work because there's a voltage across their membrane, which is set up by ion pumps and ion channels. It's a change in this voltage that transmits an action potential.

Nervous System — Synaptic Transmission

When an action potential arrives at the end of a neurone the information has to be passed on to the next cell — this could be another neurone, a muscle cell or a gland cell.

A *Synapse* is a *Junction* Between a *Neurone* and the *Next Cell*

1) A **synapse** is the junction between a **neurone** and another **neurone**, or between a **neurone** and an **effector cell**, e.g. a muscle or gland cell.

2) The **tiny gap** between the cells at a synapse is called the **synaptic cleft**.

3) The **presynaptic neurone** (the one before the synapse) has a **swelling** called a **synaptic knob**. This contains **synaptic vesicles** filled with **chemicals** called **neurotransmitters**.

4) When an **action potential** reaches the end of a neurone it causes **neurotransmitters** to be **released** into the synaptic cleft. They **diffuse across** to the **postsynaptic membrane** (the one after the synapse) and **bind** to **specific receptors**.

5) When neurotransmitters bind to receptors they might **trigger** an **action potential** (in a neurone), cause **muscle contraction** (in a muscle cell), or cause a **hormone** to be **secreted** (from a gland cell).

6) Because the receptors are **only** on the postsynaptic membranes, synapses make sure **impulses** are **unidirectional** — the impulse can only travel in **one direction**.

7) Neurotransmitters are **removed** from the **cleft** so the **response** doesn't keep happening, e.g. they're taken back into the **presynaptic neurone** or they're **broken down** by **enzymes** (and the products are taken into the neurone).

8) There are many **different** neurotransmitters. You need to know about one called **acetylcholine** (**ACh**), which binds to **cholinergic receptors**. Synapses that use acetylcholine are called **cholinergic synapses**.

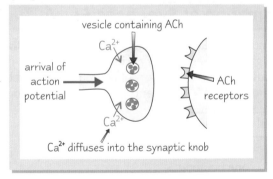

Typical structure of a synapse

ACh Transmits the Nerve Impulse *Across* a *Cholinergic Synapse*

This is how a **nerve impulse** is transmitted across a **cholinergic synapse**:

1) An action potential (see p. 54) arrives at the **synaptic knob** of the **presynaptic neurone**.

2) The action potential stimulates **voltage-gated calcium ion channels** in the **presynaptic neurone** to **open**.

3) **Calcium ions diffuse into** the synaptic knob. (They're pumped out afterwards by active transport.)

4) The influx of **calcium ions** into the synaptic knob causes the **synaptic vesicles** to **fuse** with the **presynaptic membrane**.

5) The **vesicles release** the neurotransmitter **acetylcholine** (**ACh**) into the **synaptic cleft** — this is called **exocytosis**.

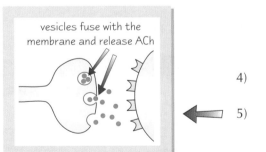

6) ACh **diffuses** across the **synaptic cleft** and **binds** to specific **cholinergic receptors** on the **postsynaptic membrane**.

7) This causes **sodium ion channels** in the **postsynaptic neurone** to **open**.

8) The **influx** of **sodium ions** into the postsynaptic membrane causes an **action potential** on the **postsynaptic membrane** (if the **threshold** is reached).

9) ACh is **removed** from the **synaptic cleft** so the **response** doesn't keep happening. It's **broken down** by an **enzyme** called **acetylcholinesterase** (**AChE**) and the products are **re-absorbed** by the **presynaptic neurone** and used to make more ACh.

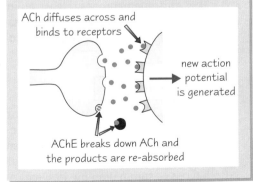

Nervous System — Synaptic Transmission

Neuromuscular Junctions are Synapses Between Neurones and Muscles

1) A **neuromuscular junction** is a **synapse** between a **motor neurone** and a **muscle cell**.

2) Neuromuscular junctions use the neurotransmitter **acetylcholine (ACh)**, which binds to cholinergic receptors called **nicotinic cholinergic receptors**.

3) Neuromuscular junctions **work** in the **same way** as the **cholinergic synapse** shown on the previous page — but there are a few **differences**:

   • The postsynaptic membrane has lots of **folds** that form **clefts**. These clefts **store** the **enzyme** that breaks down **ACh** (**acetylcholinesterase — AChE**).

   • The postsynaptic membrane has **more receptors** than other synapses.

   • When a **motor neurone** fires an **action potential**, it **always triggers** a **response** in a muscle cell. This **isn't** always the case for a synapse between two neurones (see below).

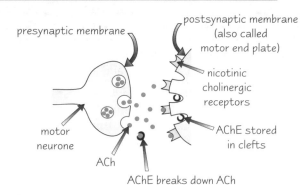

presynaptic membrane

postsynaptic membrane (also called motor end plate)

nicotinic cholinergic receptors

AChE stored in clefts

motor neurone

ACh

AChE breaks down ACh

Neurotransmitters are Excitatory or Inhibitory

1) **Excitatory** neurotransmitters **depolarise** the postsynaptic membrane, making it fire an **action potential** if the **threshold** is reached. E.g. **acetylcholine** is an excitatory neurotransmitter — it binds to cholinergic receptors to cause an **action potential** in the postsynaptic membrane.

2) **Inhibitory** neurotransmitters **hyperpolarise** the postsynaptic membrane (make the potential difference more negative), **preventing** it from firing an action potential. E.g. **GABA** is an inhibitory neurotransmitter — when it binds to its receptors it causes **potassium ion channels** to open on the postsynaptic membrane, **hyperpolarising** the neurone.

Mum couldn't help wishing Johnny had a few more inhibitory neurotransmitter

Summation at Synapses Finely Tunes the Nervous Response

If a stimulus is **weak**, only a **small amount** of **neurotransmitter** will be released from a neurone into the synaptic cleft. This might not be enough to **excite** the postsynaptic membrane to the **threshold** level and stimulate an action potential. **Summation** is where the effect of neurotransmitter released from many neurones (or one neurone that's stimulated a lot in a short period of time) is **added together**. There are two types of summation:

Spatial summation

1) Sometimes **many** neurones **connect** to **one** neurone.

2) The small amount of **neurotransmitter** released from **each** of these neurones can be enough **altogether** to **reach** the **threshold** in the postsynaptic neurone and **trigger** an **action potential**.

3) If some neurones release an **inhibitory neurotransmitter** then the total effect of all the neurotransmitters might be **no action potential**.

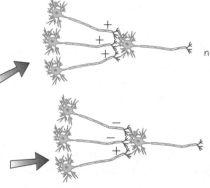

Many neurones release excitatory neurotransmitters (+) = action potential

More inhibitory neurotransmitters are released (-) than excitatory neurotransmitters (+) = no action potential

Temporal summation

Temporal summation is where **two or more** nerve impulses arrive in **quick succession** from the **same presynaptic neurone**. This makes an action potential **more likely** because **more neurotransmitter** is released into the **synaptic cleft**.

High frequency of weak excitatory inputs = action potential

Both types of **summation** mean synapses **accurately process information**, **finely tuning** the response.

Nervous System — Synaptic Transmission

Drugs Affect the Action of Neurotransmitters at Synapses in Various Ways

Some **drugs affect synaptic transmission**. You might have to **predict** the **effects** that a drug would have at a synapse in your exam. Here are some **examples** of how drugs can affect synaptic transmission:

1 Some drugs are the **same shape** as neurotransmitters so they **mimic** their action at receptors (these drugs are called **agonists**). This means **more receptors** are **activated**. E.g. **nicotine** mimics **acetylcholine** so binds to nicotinic cholinergic receptors in the brain.

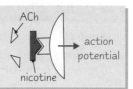
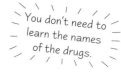

You don't need to learn the names of the drugs.

2 Some drugs **block receptors** so they **can't be activated** by neurotransmitters (these drugs are called **antagonists**). This means **fewer receptors** (if any) can be **activated**. E.g. **curare** blocks the effects of acetylcholine by blocking nicotinic cholinergic receptors at neuromuscular junctions, so muscle cells can't be stimulated. This results in the muscle being **paralysed**.

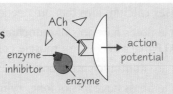

3 Some drugs **inhibit** the **enzyme** that breaks down neurotransmitters (they stop it from working). This means there are **more neurotransmitters** in the synaptic cleft to **bind** to **receptors** and they're there for **longer**. E.g. **nerve gases** stop acetylcholine from being broken down in the synaptic cleft. This can lead to **loss** of **muscle control**.

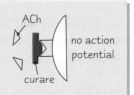

4 Some drugs **stimulate** the release of **neurotransmitter** from the presynaptic neurone so **more receptors** are activated, e.g. **amphetamines**.

5 Some drugs **inhibit** the release of neurotransmitters from the presynaptic neurone so **fewer receptors** are activated, e.g. **alcohol**.

Practice Questions

Q1 What neurotransmitter do you find at cholinergic synapses?
Q2 How do synapses ensure that nerve impulses are unidirectional?
Q3 Why are calcium ions important in synaptic transmission?
Q4 Give one way that neurotransmitters are removed from the synaptic cleft.
Q5 What kind of receptors are found at neuromuscular junctions?
Q6 What do inhibitory neurotransmitters do at synapses?
Q7 Name the two types of summation that can occur at synapses.

These questions cover pages 57-59.

Exam Questions

Q1 Describe the sequence of events from the arrival of an action potential at the presynaptic membrane of a cholinergic synapse to the generation of a new action potential at the postsynaptic membrane. [6 marks]

Q2 Myasthenia gravis is a disease in which the body's immune system gradually destroys receptors at neuromuscular junctions. Suggest what symptoms a sufferer might have. Explain your answer. [4 marks]

Q3 Galantamine is a drug that inhibits the enzyme acetylcholinesterase (AChE). Predict the effect of galantamine at a neuromuscular junction and explain your answer. [3 marks]

Neurotransmitter revision inhibits any excitement...

Another three tough pages in a row, aren't I kind to you. And lots more diagrams to have a go at drawing and re-drawing. Don't worry if you're not the world's best artist, just make sure you add labels to your drawings to explain what's happening.

Effectors — Muscle Contraction

I reckon muscle cells are definitely the spoilt brats of the biology world. They're so special that everything muscly has to have its own special name — there's none of this "cell membrane" malarkey, oh no, it's "sarcolemma" if you please... So first get your head round all these silly posh names, and then you can concentrate on what's actually going on here.

Skeletal Muscle is made up of Long Muscle Fibres

Muscles are **stimulated** to **contract** by neurones and act as **effectors**.

1) **Skeletal muscle** (also called striated, striped or voluntary muscle) is the type of muscle you use to **move**, e.g. the biceps and triceps move the lower arm.

2) Skeletal muscle is made up of **large bundles** of **long cells**, called **muscle fibres**.

3) The cell membrane of muscle fibre cells is called the **sarcolemma**.

4) Bits of the sarcolemma **fold inwards** across the muscle fibre and stick into the **sarcoplasm** (a muscle cell's cytoplasm). These folds are called **transverse (T) tubules** and they help to **spread electrical impulses** throughout the sarcoplasm so they **reach** all parts of the **muscle fibre**.

5) A network of **internal membranes** called the **sarcoplasmic reticulum** runs through the sarcoplasm. The sarcoplasmic reticulum **stores** and **releases calcium ions** that are needed for muscle contraction (see p. 62).

6) Muscle fibres have lots of **mitochondria** to **provide** the **ATP** that's needed for **muscle contraction**.

7) They are **multinucleate** (contain many nuclei).

8) Muscle fibres have lots of **long, cylindrical organelles** called **myofibrils**. They're made up of proteins and are **highly specialised** for **contraction**.

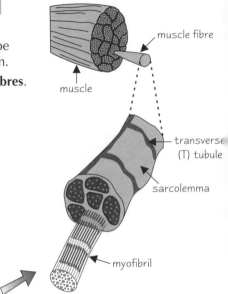

muscle fibre

muscle

transverse (T) tubule

sarcolemma

myofibril

Myofibrils Contain Thick Myosin Filaments and Thin Actin Filaments

1) Myofibrils contain bundles of **thick** and **thin myofilaments** that **move past each other** to make muscles **contract**.

- **Thick myofilaments** are made of the protein **myosin**.
- **Thin myofilaments** are made of the protein **actin**.

There's more detail on actin and myosin on p. 62.

2) If you look at a **myofibril** under an **electron microscope**, you'll see a pattern of alternating **dark** and **light bands**:

- **D**ark bands contain the **thick myosin filaments** and some overlapping thin actin filaments — these are called **A-bands**.

- **L**ight bands contain **thin actin filaments** only — these are called **I-bands**.

3) A myofibril is made up of many short units called **sarcomeres**.

4) The **ends** of each **sarcomere** are marked with a **Z-line**.

5) In the **middle** of each sarcomere is an **M-line**. The **M**-line is the **middle** of the **myosin** filaments.

6) **Around** the M-line is the **H-zone**. The H-zone **only** contains **myosin** filaments.

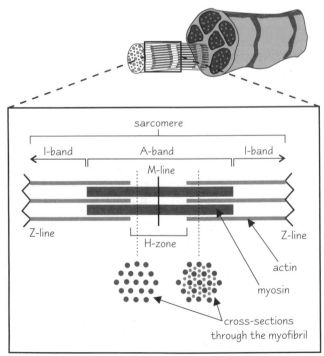

sarcomere

I-band | A-band | I-band

M-line

Z-line

H-zone

Z-line

actin

myosin

cross-sections through the myofibril

Derek was the proud winner of the biggest muscles AND the smallest pants.

Effectors — Muscle Contraction

Muscle Contraction is Explained by the Sliding Filament Theory

1) **Myosin** and **actin** filaments **slide** over one another to make the **sarcomeres contract** — the myofilaments themselves **don't** contract.

2) The **simultaneous contraction** of lots of **sarcomeres** means the **myofibrils** and **muscle fibres contract**.

3) Sarcomeres return to their **original length** as the muscle **relaxes**.

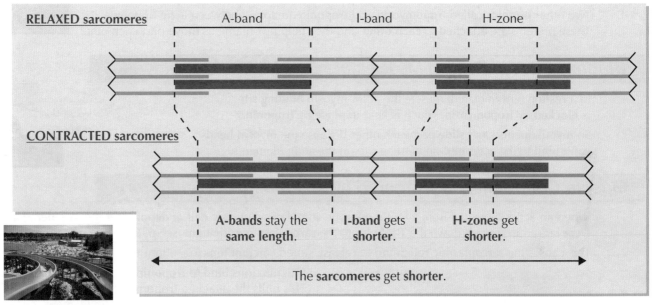

If only the sliding filament theory was as much fun...

Practice Questions

Q1 What are transverse (T) tubules?

Q2 Name the two proteins that make up myofibrils.

Q3 What are the dark bands called?

Q4 What are the light bands called?

Q5 How do myofilaments make sarcomeres contract?

Q6 What happens to sarcomeres as a muscle relaxes?

Exam Questions

Q1 Describe how myofilaments, muscle fibres, myofibrils and muscles are related to each other. [3 marks]

Q2 A muscle myofibril was examined under an electron microscope and a sketch was drawn (Figure 1).

a) What are the correct names for labels A, B and C? [3 marks]

b) Describe how the lengths of the different bands in a myofibril change during muscle contraction. [2 marks]

c) The myofibril was then cut through the M-line (Figure 2). State which of the cross-section drawings you would expect to see and explain why. [3 marks]

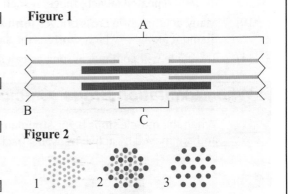

Figure 1

Figure 2

Sarcomere — a French mother with a dry sense of humour...

Blimey, there are an awful lot of similar-sounding names to learn on these pages. And then you've got your A-band, I-band, what-band, who-band to memorise too. But once you've learnt them, these are things you'll never forget — that's right, they'll take up vital brain space forever. And they'll also get you vital marks in your exam.

Effectors — Muscle Contraction

Myofilaments sliding over one another takes a lot of energy — probably why exercise is such hard work...

Myosin Filaments Have Globular Heads and Binding Sites

1) **Myosin filaments** have **globular heads** that are **hinged**, so they can move **back** and **forth**.
2) Each myosin head has a **binding site** for **actin** and a **binding site** for **ATP**.
3) **Actin filaments** have **binding sites** for **myosin heads**, called **actin-myosin** binding sites.
4) Two other **proteins** called **tropomyosin** and **troponin** are found between actin filaments. These proteins are **attached** to **each other** and they **help** myofilaments **move** past each other.

Binding Sites in Resting Muscles are Blocked by Tropomyosin

1) In a **resting** (unstimulated) muscle the **actin-myosin binding site** is **blocked** by **tropomyosin**, which is held in place by **troponin**.
2) So **myofilaments can't slide** past each other because the **myosin heads can't bind** to the actin-myosin binding site on the actin filaments.

Muscle Contraction is Triggered by an Influx of Calcium Ions

1) When an action potential from a motor neurone **stimulates** a muscle cell, it **depolarises** the **sarcolemma**. Depolarisation **spreads** down the **T-tubules** to the **sarcoplasmic reticulum** (see p. 60).
2) This causes the **sarcoplasmic reticulum** to **release** stored **calcium ions** (Ca^{2+}) into the **sarcoplasm**.

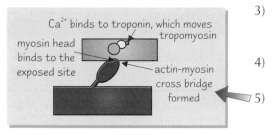

3) Calcium ions **bind** to **troponin**, causing it to **change shape**. This **pulls** the attached **tropomyosin out** of the **actin-myosin binding site** on the actin filament.
4) This **exposes** the **binding site**, which allows the **myosin head** to **bind**.
5) The bond formed when a **myosin head** binds to an **actin filament** is called an **actin-myosin cross bridge**.

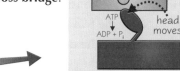

6) **Calcium** ions also **activate** the enzyme **ATPase**, which **breaks down ATP** (into ADP + P$_i$) to **provide** the **energy** needed for muscle contraction.
7) The **energy** released from ATP **moves** the **myosin head**, which **pulls** the **actin filament** along in a kind of **rowing action**.
8) **ATP** also provides the **energy** to **break** the **actin-myosin cross bridge**, so the **myosin head detaches** from the actin filament **after** it's moved.
9) The **myosin head** then **reattaches** to a **different binding site** further along the actin filament. A **new actin-myosin cross bridge** is formed and the **cycle** is **repeated** (attach, move, detach, reattach to new binding site...).

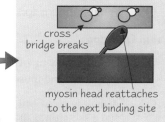

10) **Many** actin-myosin cross bridges **form** and **break** very **rapidly**, pulling the actin filament along — which **shortens** the **sarcomere**, causing the **muscle** to **contract**.
11) The cycle will **continue** as long as **calcium ions** are **present** and **bound** to **troponin**.

When Excitation Stops, Calcium Ions Leave Troponin Molecules

1) When the muscle **stops** being **stimulated**, **calcium ions leave** their **binding sites** on the **troponin** molecules and are moved by **active transport** back into the **sarcoplasmic reticulum** (this needs **ATP** too).

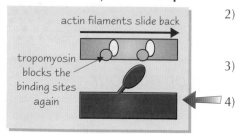

2) The **troponin** molecules return to their **original shape**, pulling the attached **tropomyosin** molecules with them. This means the **tropomyosin** molecules **block** the actin-myosin **binding sites** again.
3) Muscles **aren't contracted** because **no myosin heads** are **attached** to **actin** filaments (so there are no actin-myosin cross bridges).
4) The **actin** filaments **slide back** to their **relaxed** position, which **lengthens** the **sarcomere**.

Effectors — Muscle Contraction

ATP and Phosphocreatine Provide the Energy for Muscle Contraction

So much **energy** is **needed** when muscles contract that **ATP** gets **used up very quickly**.
ATP has to be **continually generated** so exercise can continue — this happens in **three main ways**:

1) <u>Aerobic respiration</u>
   - Most **ATP** is generated via **oxidative phosphorylation** in the cell's **mitochondria**.
   - **Aerobic** respiration only works when there's **oxygen** so it's good for **long periods** of **low-intensity exercise**.

2) <u>Anaerobic respiration</u>

   See pages 22-25 for more on aerobic and anaerobic respiration.

   - ATP is made **rapidly** by **glycolysis**.
   - The **end product** of glycolysis is **pyruvate**, which is converted to **lactate** by **lactate fermentation**.
   - Lactate can **quickly build up** in the muscles and cause **muscle fatigue**.
   - Anaerobic respiration is good for **short periods** of **hard exercise**, e.g. a **400 m sprint**.

3) <u>ATP-Phosphocreatine (PCr) System</u>
   - ATP is made by **phosphorylating ADP** — adding a phosphate group taken from **PCr**.

   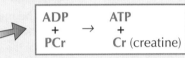

   $$\begin{array}{ccc} \text{ADP} & & \text{ATP} \\ + & \rightarrow & + \\ \text{PCr} & & \text{Cr (creatine)} \end{array}$$

   Many activities use a combination of these systems.

   - PCr is **stored** inside cells and the ATP-PCr system **generates ATP** very **quickly**.
   - **PCr runs out** after a few seconds so it's used during **short bursts** of **vigorous exercise**, e.g. a **tennis serve**.
   - The ATP-PCr system is **anaerobic** (it doesn't need oxygen) and it's **alactic** (it doesn't form any lactate).

Skeletal Muscles are Made of Slow Twitch and Fast Twitch Muscle Fibres

Skeletal muscles are made up of **two types** of **muscle fibres** — **slow twitch** and **fast twitch**.
Different muscles have **different proportions** of slow and fast twitch fibres. The two types have **different properties**:

SLOW TWITCH MUSCLE FIBRES	FAST TWITCH MUSCLE FIBRES
Muscle fibres that contract slowly.	Muscle fibres that contract very quickly.
Muscles you use for posture, e.g. those in the back, have a high proportion of them.	Muscles you use for fast movement, e.g. those in the eyes and legs, have a high proportion of them.
Good for endurance activities, e.g. maintaining posture, long-distance running.	Good for short bursts of speed and power, e.g. eye movement, sprinting.
Can work for a long time without getting tired.	Get tired very quickly.
Energy's released slowly through aerobic respiration. Lots of mitochondria and blood vessels supply the muscles with oxygen.	Energy's released quickly through anaerobic respiration using glycogen (stored glucose). There are few mitochondria or blood vessels.
Reddish in colour because they're rich in myoglobin — a red-coloured protein that stores oxygen.	Whitish in colour because they don't have much myoglobin (so can't store much oxygen).

Practice Questions

Q1 Describe one way that ATP can be generated in contracting muscles.

Q2 State three differences between slow and fast twitch skeletal muscle fibres.

Exam Questions

Q1 Rigor mortis is the stiffening of muscles in the body after death. It happens when ATP reserves are exhausted. Explain why a lack of ATP leads to muscles being unable to relax. [3 marks]

Q2 Bepridil is a drug that blocks calcium ion channels.
Describe and explain the effect this drug will have on muscle contraction. [3 marks]

What does muscle contraction cost? 80p...

*Sorry, that's my favourite sciencey joke so I had to fit it in somewhere — a small distraction before you revisit this page.
It's tough stuff but you know the best way to learn it. That's right, grab yourself a nice felt-tip pen and a pad of paper...*

Responses in Animals

Right, that's enough pages on receptors, effectors and whatnot — here are some real-life examples to bring it all together.

Control of Heart Rate Involves the Brain and Autonomic Nervous System

1) The **sinoatrial node** (**SAN**) generates **electrical impulses** that cause the **cardiac muscles** to **contract**.

2) The **rate** at which the SAN fires (i.e. heart rate) is **unconsciously controlled** by a part of the **brain** called the **medulla**.

3) Animals need to **alter** their **heart rate** to **respond** to **internal stimuli**, e.g. to prevent fainting due to low blood pressure or to make sure the heart rate is high enough to supply the body with enough oxygen.

4) **Stimuli** are **detected** by **pressure receptors** and **chemical receptors**:
   - There are **pressure receptors** called **baroreceptors** in the **aorta** and the **vena cava**. They're stimulated by **high** and **low blood pressure**.
   - There are **chemical receptors** called **chemoreceptors** in the **aorta**, the **carotid artery** (a major artery in the neck) and in the **medulla**. They **monitor** the **oxygen** level in the **blood** and also **carbon dioxide** and **pH** (which are indicators of O_2 level).

   There's more about the autonomic nervous system on page 50.

5) Electrical impulses from receptors are sent **to the medulla** along **sensory** neurones. The medulla processes the information and sends impulses to the SAN along **sympathetic** or **parasympathetic** neurones (which are part of the **autonomic nervous system**). Here's how it all works:

Stimulus	Receptor	Neurone and neurotransmitter	Effector	Response
High blood pressure.	Baroreceptors detect high blood pressure.	Impulses are sent to the medulla, which sends impulses along parasympathetic neurones. These secrete acetylcholine, which binds to receptors on the SAN.	Cardiac muscles	Heart rate slows down to reduce blood pressure back to normal.
Low blood pressure.	Baroreceptors detect low blood pressure.	Impulses are sent to the medulla, which sends impulses along sympathetic neurones. These secrete noradrenaline (a neurotransmitter), which binds to receptors on the SAN.	Cardiac muscles	Heart rate speeds up to increase blood pressure back to normal.
High blood O_2, low CO_2 or high pH levels.	Chemoreceptors detect chemical changes in the blood.	Impulses are sent to the medulla, which sends impulses along parasympathetic neurones. These secrete acetylcholine, which binds to receptors on the SAN.	Cardiac muscles	Heart rate decreases to return O_2, CO_2 and pH levels back to normal.
Low blood O_2, high CO_2 or low pH levels.	Chemoreceptors detect chemical changes in the blood.	Impulses are sent to the medulla, which sends impulses along sympathetic neurones. These secrete noradrenaline, which binds to receptors on the SAN.	Cardiac muscles	Heart rate increases to return O_2, CO_2 and pH levels back to normal.

Reflexes are Rapid, Automatic Responses to Stimuli

1) A **reflex** is where the body **responds** to a stimulus **without** making a **conscious decision** to respond.

2) Because you don't have to **spend time deciding** how to respond, information travels **really fast** from **receptors** to **effectors**.

3) So simple reflexes help organisms to **avoid damage** to the body because they're **rapid**.

4) The **pathway** of neurones linking receptors to effectors in a reflex is called a **reflex arc**. You need to **learn** a **simple reflex arc** involving three neurones — a **sensory**, a **relay** and a **motor** neurone.

> ### E.g. the **hand-withdrawal response** to heat
>
> - **Thermoreceptors** in the skin **detect** the **heat** stimulus.
> - The **sensory neurone** carries impulses to the **relay neurone**.
> - The **relay neurone** connects to the **motor neurone**.
> - The **motor neurone** sends **impulses** to the **effector** (your biceps muscle).
> - Your **muscle contracts** to **stop** your hand being **damaged**. ⟹

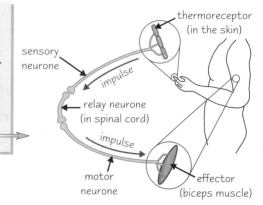

5) If there's a **relay neurone** involved in the simple reflex arc then it's possible to **override** the reflex, e.g. in the example above your **brain** could tell your hand to **withstand** the heat.

Responses in Animals

Simple Responses Keep Simple Organisms in a Favourable Environment

Simple organisms, e.g. woodlice and earthworms, have **simple responses** to keep them in a **favourable environment**. Their **response** can either be **tactic** or **kinetic**:

- **Tactic responses** (**taxes**) — the organisms move towards or away from a **directional stimulus**.

  > For example, **woodlice** show a **tactic** response to light (**phototaxis**) — they move **away from** a **light source**. This helps them **survive** as it keeps them **concealed** under stones during the day (where they're **safe** from predators) and keeps them in **damp conditions** (which reduces water loss).

- **Kinetic responses** (**kineses**) — the organisms' movement is affected by a **non-directional** stimulus, e.g. **intensity**.

  > For example, **woodlice** show a **kinetic** response to **humidity**. In **high humidity** they move **slowly** and **turn** more often, so that they **stay where they are**. As the air gets **drier**, they move **faster** and **turn less** often, so that they move into a **new area**. This response helps woodlice **move** from **drier air** to more **humid air**, and then **stay put**. This **improves** the **survival** chances of the organism — it **reduces** their **water loss** and it helps to keep them **concealed**.

Some Cells Communicate with Other Cells by Secreting Chemical Mediators

1) A **chemical mediator** is a **chemical messenger** that acts **locally** (i.e. on **nearby cells**).

2) Communication using chemical mediators is **similar** to communication using **hormones** (see p. 51) — cells release chemicals that bind to **specific receptors** on **target cells** to cause a **response**. But there are a few **differences**:

   - Chemical mediators are secreted from **cells** that are **all over** the **body** (not just from glands).
   - Their **target cells** are right **next to** where the chemical mediator's produced. This means they stimulate a **local response** (not a widespread one).
   - They only have to travel a **short distance** to their target cells, so produce a **quicker response** than hormones (which are transported in the blood).

3) You need to **know** about **two** types of **chemical mediator** — histamine and prostaglandins:

HISTAMINE	**Histamine** is a chemical mediator that's stored in **mast cells** and **basophils** (types of immune system cell). It's **released** in response to the body being **injured** or **infected**. It increases the **permeability** of the **capillaries nearby** to allow more immune system cells to move out of the blood to the infected or injured area.
PROSTAGLANDINS	**Prostaglandins** are a **group** of chemical mediators that are produced by **most cells** of the **body**. They're involved in loads of things like **inflammation**, **fever**, **blood pressure regulation** and **blood clotting**. E.g. one type of prostaglandin is released from blood vessel epithelium cells and causes the muscles around them to relax.

Practice Questions

Q1 Describe a simple reflex arc.
Q2 What's the difference between taxes and kineses?
Q3 What is a chemical mediator?

Exam Question

Q1 a) Explain how high blood pressure in the aorta causes the heart rate to slow down. [5 marks]

b) What would be the effect of severing the nerves from the medulla to the sinoatrial node (SAN)? [2 marks]

AAAAAAAAAAAAAAAAAAAAAAAARGH — the reflex response to revision...

There's also the good old tactic response to revision — when you see your revision notes, you always move away from them. I suppose that wouldn't really help, unless you put your notes miles apart from each other and used it as a fitness exercise. Better still, get on and learn this stuff because you've nearly finished this section — bet that stimulates an excitable response.

Survival and Responses in Plants

Plants also have ways of responding to stimuli — OK so they're not as quick as animals, but they're important all the sam

Plants Need to Respond to Stimuli Too

Flowering plants, like animals, **increase** their chances of **survival** by **responding** to changes in their **environment**, e.g:
* They sense the direction of **light** and **grow** towards it to **maximise** light absorption for **photosynthesis**.
* They can sense **gravity**, so their roots and shoots **grow** in the **right direction**.
* **Climbing** plants have a sense of **touch**, so they can find things to climb and **reach** the **sunlight**.

A Tropism is a Plant's Growth Response to an External Stimulus

1) A **tropism** is the **response** of a plant to a **directional stimulus** (a stimulus coming from a particular direction).
2) Plants respond to stimuli by **regulating** their **growth**.
3) A **positive** tropism is growth **towards** the stimulus.
4) A **negative** tropism is growth **away** from the stimulus.

* **Phototropism** is the growth of a plant in response to **light**.
* **Shoots** are **positively phototropic** and grow **towards** light.
* **Roots** are **negatively phototropic** and grow **away** from light.

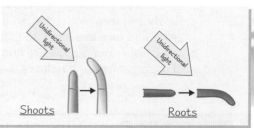

Shoots   Roots

* **Geotropism** is the growth of a plant in response to **gravity**.
* **Shoots** are **negatively geotropic** and grow **upwards**.
* **Roots** are **positively geotropic** and grow **downwards**.

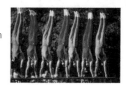

The men's gymnastics team were negatively geotropic.

Responses are Brought About by Growth Factors

1) Plants **don't** have a **nervous system** so they can't respond using neurones, and they **don't** have a **circulatory system** so they can't respond using hormones either.
2) Plants **respond** to stimuli using **growth factors** — these are chemicals that **speed up** or **slow down** plant **growth**.
3) Growth factors are **produced** in the **growing regions** of the plant (e.g. shoot tips, leaves) and they **move** to where they're needed in the **other parts** of the plant.
4) A growth factor called **gibberellin** stimulates **flowering** and **seed germination**.
5) Growth factors called **auxins** stimulate the **growth** of shoots by **cell elongation** — this is where **cell walls** become **loose** and **stretchy**, so the cells get **longer**.
6) **High** concentrations of auxins **inhibit growth** in **roots** though.

Indoleacetic Acid (IAA) is an Important Auxin

1) **Indoleacetic acid** (**IAA**) is an important **auxin** that's produced in the **tips** of **shoots** in flowering plants.
2) IAA is **moved** around the plant to **control tropisms** — it moves by **diffusion** and **active transport** over short distances, and via the **phloem** over long distances.
3) This results in **different parts** of the plants having **different amounts** of IAA. The **uneven distribution** of IAA means there's **uneven growth** of the plant, e.g:

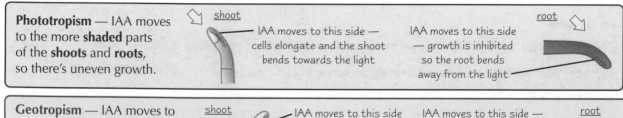

Phototropism — IAA moves to the more **shaded** parts of the **shoots** and **roots**, so there's uneven growth.

shoot

IAA moves to this side — cells elongate and the shoot bends towards the light

IAA moves to this side — growth is inhibited so the root bends away from the light

root

Geotropism — IAA moves to the **underside** of **shoots** and **roots**, so there's uneven growth.

shoot

IAA moves to this side — cells elongate so the shoot grows upwards

IAA moves to this side — growth is inhibited so the root grows downwards

root

Survival and Responses in Plants

You May Have to **Interpret Experimental Data** About **Auxins**

Here's some **data** similar to what you might get in your **exam**:

1) An experiment was carried out to **investigate** the role of **auxin** in **shoot growth**.

2) Eight shoots, **equal in height and mass**, had their **tips removed**.

3) **Sponges** soaked in **glucose and either auxin** or **water** were then ⟹ placed where the tip should be.

4) **Four tips** were then placed in the **dark** (experiment A) and the **other** four tips were exposed to a **light** source, directed at them from the **right** (experiment B).

5) After **two days** the **amount** of growth (in mm) and **direction** of growth was **recorded**. The results are shown in the table. ⟹

■ Sponge soaked in auxin and glucose
■ Sponge soaked in water and glucose

A B C D ← Shoot minus the tip

	Growth			
	Shoot A	Shoot B	Shoot C	Shoot D
Experiment A (dark)	6 mm, right	6 mm, left	6 mm, straight	1 mm, straight
Experiment B (light)	8 mm, right	8 mm, right	8 mm, right	3 mm, right

You could be asked to **explain the data**:

The results show how the **movement** of **auxin** controls **phototropism** in plant shoots.

In **experiment A shoot A**, the auxin **diffused** straight down from the sponge into the **left-hand side** of the shoot. This stimulated the cells on this side to **elongate**, so the shoot **grew towards the right**. In shoot B, the opposite occurred, making the shoot **grow towards the left**. In shoot C, **equal amounts** of auxin diffused down **both sides**, making **all** the cells elongate at the same rate.

In **experiment B**, the shoots were exposed to a **light source**. The auxin diffused into the shoot and **accumulated on the shaded side** (left-hand side) **regardless** of **where** the sponge was placed. All the shoots **grew towards the right**, because most auxin **accumulated** on the left, **stimulating** cell elongation there.

You could be asked to comment on the **experimental design**:

A **control** (sponge soaked in water) was included to show that it was the auxin having an **effect** and nothing else. **Glucose** was included so that the shoots would have **energy to grow in the dark** (no photosynthesis can take place).

There's more on pages 106-108 about interpreting experimental data.

Practice Questions

Q1 What is a tropism?

Q2 What are plant growth factors?

Q3 How does the movement of IAA control geotropism in roots?

Week	Height of plant not given auxins / cm	Height of plant provided with auxins / cm
1	1	2
2	2	5
3	4	8
4	6	9
5	9	13

Exam Question

Q1 The table shows the results some students obtained when they investigated the effect of providing plants with auxins.

a) Describe and explain what the data shows. [2 marks]

b) Explain the role of auxins in the control of phototropism. [5 marks]

c) Suggest what the students should do to increase the reliability of their results. [2 marks]

d) Why might this data be useful to a commercial tomato producer? [1 mark]

IAA Productions — do you have the growth factor — with Simon Trowel...

Hoorah, you've finally reached the end of this mega-section. Unless you're casually skipping through the book, in which case you've got the treat of all treats waiting for you. There are just a few more gibberish names (sorry, I mean gibberellin) for you to learn, and some rather bendy responses to understand, then you've got responses all sorted. Aux-ilarating.

Homeostasis Basics

Ah, there's nothing like learning a nice long word to start you off on a new section — welcome to homeostasis.

Homeostasis *is the* Maintenance *of a* Constant Internal Environment

1) **Changes** in your **external environment** can affect your **internal environment**
   — the blood and tissue fluid that surrounds your cells.

2) **Homeostasis** involves **control systems** that keep your **internal environment** roughly **constant** (within **certain limits**).

3) **Keeping** your internal environment **constant** is vital for cells to **function normally** and to **stop** them being **damaged**.

4) It's particularly important to **maintain** the right **core body temperature** and **blood pH**.
   This is because temperature and pH affect **enzyme activity**, and enzymes **control** the **rate** of **metabolic reactions**:

   Temperature
   - If body temperature is **too high** (e.g. 40 °C) **enzymes** may become **denatured**. The enzyme's molecules **vibrate too much**, which **breaks** the **hydrogen bonds** that hold them in their **3D shape**. The **shape** of the enzyme's **active site** is **changed** and it **no longer works** as a **catalyst**. This means **metabolic reactions** are **less efficient**.
   - If body temperature is **too low enzyme activity** is **reduced, slowing** the rate of **metabolic reactions**.
   - The **highest rate** of **enzyme activity** happens at their **optimum temperature** (about **37 °C** in humans).

   pH
   - If blood pH is **too high** or **too low** (highly alkaline or acidic) **enzymes** become **denatured**. The **hydrogen bonds** that hold them in their 3D shape are **affected** so the **shape** of the enzyme's **active site** is **changed** and it **no longer works** as a **catalyst**. This means **metabolic reactions** are **less efficient**.
   - The **highest rate** of **enzyme activity** happens at their **optimum pH** — usually **around pH 7** (neutral), but some enzymes work best at other pHs, e.g. enzymes found in the stomach work best at a low pH.

5) It's important to **maintain** the right **concentration** of **glucose** in the **blood** because cells need glucose for **energy**.
   Blood glucose concentration also affects the **water potential** of blood — this is the potential (likelihood) of water molecules to **diffuse** out of or into a solution.

   Glucose
   - If blood glucose concentration is **too high** the **water potential** of blood is **reduced** to a point where **water** molecules **diffuse out** of cells into the blood by osmosis. This can cause the cells to **shrivel up** and **die**.
   - If blood glucose concentration is **too low**, cells are **unable** to carry out **normal activities** because there **isn't enough glucose** for respiration to provide **energy**.

Homeostatic Systems *Detect a* Change *and* Respond *by* Negative Feedback

1) Homeostatic systems involve **receptors**, a **communication system** and **effectors** (see page 50).

2) Receptors detect when a level is **too high** or **too low**, and the information's communicated via the **nervous** system or the **hormonal** system to **effectors**.

3) The effectors respond to **counteract** the change — bringing the level **back** to **normal**.

4) The mechanism that **restores** the level to **normal** is called a **negative feedback** mechanism.

5) Negative feedback **keeps** things around the **normal** level, e.g. body temperature is usually kept **within 0.5 °C** above or below **37 °C**.

6) Negative feedback only works within **certain limits** though — if the change is **too big** then the **effectors** may **not** be able to **counteract** it, e.g. a huge drop in body temperature caused by prolonged exposure to cold weather may be too large to counteract.

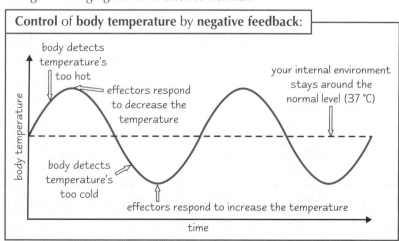

Control of **body temperature** by **negative feedback**:

body detects temperature's too hot

effectors respond to decrease the temperature

your internal environment stays around the normal level (37 °C)

body detects temperature's too cold

effectors respond to increase the temperature

body temperature

time

Homeostasis Basics

Multiple Negative Feedback Mechanisms Give More Control

1) Homeostasis involves **multiple negative feedback mechanisms** for each thing being controlled. This is because having more than one mechanism gives **more control** over changes in your internal environment than just having one negative feedback mechanism.

2) Having multiple negative feedback mechanisms means you can **actively increase** or **decrease a level** so it returns to **normal**, e.g. you have feedback mechanisms to reduce your body temperature and you also have mechanisms to increase it.

3) If you only had **one negative feedback mechanism**, all you could do would be **turn it on** or **turn it off**. You'd only be able to actively change a level in **one direction** so it returns to normal, e.g. it's a bit like trying to slow down a car with only an accelerator — all you can do is take your foot off the accelerator (you'd have more control with a brake too).

4) Only **one** negative feedback mechanism means a **slower response** and **less control**.

There was plenty of negative feedback when Carl wore his new vest-pants combo out for dinner.

Positive Feedback Mechanisms Amplify a Change from the Normal Level

1) Some changes trigger a **positive feedback** mechanism, which **amplifies** the change.

2) The effectors respond to **further increase** the level **away** from the **normal** level.

3) Positive feedback is useful to **rapidly activate** something, e.g. a **blood clot** after an injury.

- **Platelets** become **activated** and release a **chemical** — this triggers **more platelets** to be activated, and so on.
- Platelets **very quickly** form a **blood clot** at the injury site.
- The process **ends** with **negative feedback**, when the body detects the **blood clot** has been **formed**.

4) Positive feedback can also happen when a **homeostatic system breaks down**, e.g. if you're too cold for too long:

Hypothermia involves **positive feedback**:

- **Hypothermia** is **low body temperature** (below 35 °C).
- It happens when **heat's lost** from the body **quicker** than it can be **produced**.
- As body temperature **falls** the **brain doesn't work** properly and **shivering stops** — this makes body temperature **fall even more**.
- **Positive feedback** takes body temperature **further away** from the normal level, and it continues to decrease unless action is taken.

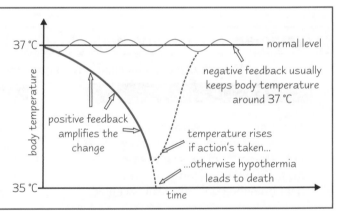

normal level

negative feedback usually keeps body temperature around 37 °C

positive feedback amplifies the change

temperature rises if action's taken...

...otherwise hypothermia leads to death

5) Positive feedback **isn't** involved in **homeostasis** because it **doesn't** keep your internal environment **constant**.

Practice Questions

Q1 What is homeostasis and why is it necessary?

Q2 Why is it important to control blood pH?

Q3 Why is it important to control blood glucose concentration?

Statement A: "Hyperthermia happens when the brain can't work properly and body temperature continues to increase."
Statement B: "When body temperature is low, mechanisms return the temperature to normal."

Exam Questions

Q1 Look at statements A and B in the box.
   a) Which statement is describing a positive feedback mechanism? Give a reason for your answer. [2 marks]
   b) Describe and explain what effect a very high body temperature has on metabolic reactions. [2 marks]

Q2 Describe the importance of multiple negative feedback mechanisms in homeostasis. [2 marks]

Homeostasis works like a teacher — everything always gets corrected...

The key to understanding homeostasis is to get your head around negative feedback. Basically, if one thing goes up, the body responds to bring it down — and vice versa. And when you're ready, turn over the page for some exciting examples.

Control of Body Temperature

So, negative feedback is a good thing — well, in biological terms anyway. Being able to control your temperature is really important and there are some pretty nifty mechanisms that help you do this. Read on, oh chosen one, read on...

Temperature is Controlled Differently in Ectotherms and Endotherms

Animals are classed as either **ectotherms** or **endotherms**, depending on how they **control** their body temperature:

Ectotherms — e.g. reptiles, fish	Endotherms — e.g. mammals, birds
Ectotherms **can't control** their body temperature **internally** — they **control** their temperature by **changing** their **behaviour** (e.g. reptiles gain heat by basking in the sun).	Endotherms **control** their body temperature **internally** by **homeostasis**. They can also control their temperature by **behaviour** (e.g. by finding shade).
Their **internal** temperature **depends** on the **external temperature** (their surroundings).	Their internal temperature is **less affected** by the **external temperature** (within certain limits).
Their **activity** level **depends** on the external temperature — they're **more** active at **higher** temperatures and **less** active at **lower** temperatures.	Their **activity** level is largely **independent** of the **external temperature** — they can be active at any temperature (within certain limits).
They have a **variable metabolic rate** and they **generate** very **little heat** themselves.	They have a constantly **high metabolic rate** and they **generate** a **lot** of **heat** from metabolic reactions.

Mammals have Many Mechanisms to Change Body Temperature

You need to **learn** the different **mechanisms** that mammals use to **change body temperature**:

Heat loss

Sweating — **more sweat** is secreted from **sweat glands** when the body's too hot. The water in sweat **evaporates** from the surface of the skin and **takes heat** from the body. The **skin is cooled**.

Hairs lie flat — mammals have a layer of **hair** that provides **insulation** by **trapping air** (air is a poor conductor of heat). When it's hot, **erector pili muscles relax** so the hairs lie flat. **Less air** is trapped, so the skin is **less insulated** and **heat** can be **lost** more easily.

Vasodilation — when it's hot, **arterioles** near the surface of the skin **dilate** (this is called **vasodilation**). **More blood** flows through the **capillaries** in the surface layers of the dermis. This means **more heat** is **lost** from the skin by **radiation** and the **temperature is lowered**.

Heat production

Shivering — when it's cold, **muscles contract** in **spasms**. This makes the body **shiver** and **more heat** is **produced** from **increased respiration**.

Hormones — the body releases **adrenaline** and **thyroxine**. These **increase metabolism** and so **more heat** is **produced**.

Heat conservation

Much less sweat — less sweat is secreted from sweat glands when it's cold, **reducing** the amount of **heat loss**.

Hairs stand up — **erector pili muscles contract** when it's cold, which makes the **hairs stand up**. This **traps more air** and so **prevents heat loss**.

Vasoconstriction — when it's cold, **arterioles** near the surface of the skin **constrict** (this is called **vasoconstriction**) so **less blood** flows through the **capillaries** in the surface layers of the dermis. This **reduces heat loss**.

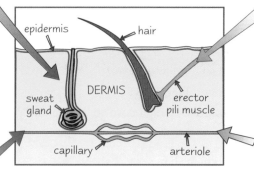

epidermis — hair — DERMIS — sweat gland — erector pili muscle — capillary — arteriole

Control of Body Temperature

The **Hypothalamus Controls** Body Temperature in **Mammals**

1) **Body temperature** in mammals is **maintained** at a **constant level** by a part of the **brain** called the **hypothalamus**.

2) The hypothalamus **receives information** about both **internal** and **external temperature** from **thermoreceptors** (temperature receptors):

Information about **internal temperature** comes from **thermoreceptors** in the **hypothalamus** that detect **blood temperature**.	Information about **external temperature** comes from **thermoreceptors** in the **skin** that detect **skin temperature**.

3) Thermoreceptors send **impulses** along **sensory neurones** to the **hypothalamus**, which sends **impulses** along **motor neurones** to **effectors** (muscles and glands).

4) The neurones are part of the **autonomic nervous system**, so it's all done **unconsciously**.

5) The effectors respond to **restore** the body temperature **back to normal**. Here's how it all works:

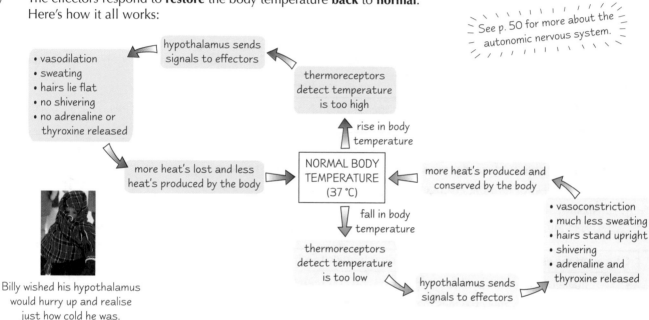

See p. 50 for more about the autonomic nervous system.

- vasodilation
- sweating
- hairs lie flat
- no shivering
- no adrenaline or thyroxine released

hypothalamus sends signals to effectors

thermoreceptors detect temperature is too high

rise in body temperature

more heat's lost and less heat's produced by the body

NORMAL BODY TEMPERATURE (37 °C)

more heat's produced and conserved by the body

fall in body temperature

thermoreceptors detect temperature is too low

hypothalamus sends signals to effectors

- vasoconstriction
- much less sweating
- hairs stand upright
- shivering
- adrenaline and thyroxine released

Billy wished his hypothalamus would hurry up and realise just how cold he was.

Practice Questions

Q1 Give four differences between ectotherms and endotherms.

Q2 Which type of animal has more control over their body temperature, ectotherms or endotherms?

Q3 How does sweating reduce body temperature?

Q4 What role do erector pili muscles play in controlling body temperature in mammals?

Q5 Which part of the brain is responsible for maintaining a constant body temperature in mammals?

Exam Questions

Q1 Mammals that live in cold climates have thick fur and layers of fat beneath their skin to keep them warm. Describe and explain two other ways they maintain a constant body temperature in cold conditions. [4 marks]

Q2 Describe and explain how the body detects a high external temperature. [2 marks]

Q3 Snakes are usually found in warm climates. Suggest why they are not usually found in cold climates. Explain your answer. [4 marks]

Sweat, hormones and erector muscles — ooooh errrrrrr...

Blimey, I'm glad this is all done unconsciously — you'd waste tons of time if you had to think about every single response. Mind you, I reckon I could think up some slightly less embarrassing ways of controlling temperature, rather than getting all red-faced and stinky. Ectotherms have got it sussed with their whole sunbathing thing — now that's the life...

Control of Blood Glucose Concentration

These pages are all about how negative feedback helps you to not go totally hyper when you stuff your face with sweets.

Eating and Exercise Change the Concentration of Glucose in your Blood

1) **All cells** need a constant **energy supply** to work — so **blood glucose concentration** must be carefully **controlled**.
2) The **concentration** of **glucose** in the blood is **normally** around **90 mg per 100 cm³** of blood.
   It's **monitored** by cells in the **pancreas**.
3) Blood glucose concentration **rises** after **eating food** containing **carbohydrate**.
4) Blood glucose concentration **falls** after **exercise**, as **more glucose** is used in **respiration** to **release energy**.

Insulin and Glucagon Control Blood Glucose Concentration

The hormonal system (see p. 51) **controls** blood glucose concentration using **two hormones** called **insulin** and **glucagon**. They're both **secreted** by clusters of cells in the **pancreas** called the **islets of Langerhans**:

- **Beta (β) cells** secrete **insulin** into the blood.
- **Alpha (α) cells** secrete **glucagon** into the blood.

Insulin and glucagon act on **effectors**, which respond to **restore** the blood glucose concentration to the **normal level**:

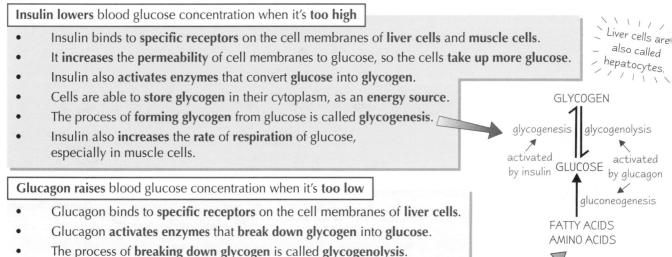

Insulin lowers blood glucose concentration when it's **too high**

- Insulin binds to **specific receptors** on the cell membranes of **liver cells** and **muscle cells**.
- It **increases** the **permeability** of cell membranes to glucose, so the cells **take up more glucose**.
- Insulin also **activates enzymes** that convert **glucose** into **glycogen**.
- Cells are able to **store glycogen** in their cytoplasm, as an **energy source**.
- The process of **forming glycogen** from glucose is called **glycogenesis**.
- Insulin also **increases** the **rate** of **respiration** of glucose, especially in muscle cells.

Liver cells are also called hepatocytes.

Glucagon raises blood glucose concentration when it's **too low**

- Glucagon binds to **specific receptors** on the cell membranes of **liver cells**.
- Glucagon **activates enzymes** that **break down glycogen** into **glucose**.
- The process of **breaking down glycogen** is called **glycogenolysis**.
- Glucagon also promotes the formation of glucose from **fatty acids** and **amino acids**.
- The process of **forming glucose** from **non-carbohydrates** is called **gluconeogenesis**.
- Glucagon **decreases** the **rate** of respiration of glucose in cells.

Negative Feedback Mechanisms Keep Blood Glucose Concentration Normal

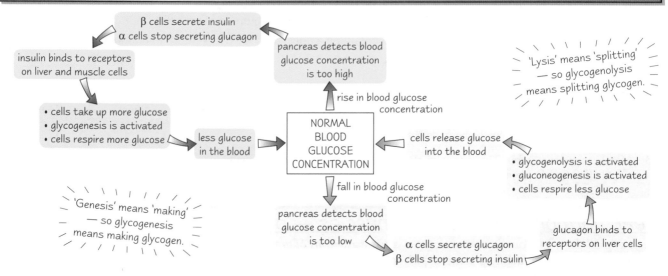

'Lysis' means 'splitting' — so glycogenolysis means splitting glycogen.

'Genesis' means 'making' — so glycogenesis means making glycogen.

Control of Blood Glucose Concentration

Adrenaline Increases your Blood Glucose Concentration Too

1) **Adrenaline** is a **hormone** that's secreted from your **adrenal glands** (found just above your kidneys).

2) It's secreted when there's a **low concentration** of **glucose** in your blood, when you're **stressed** and when you're **exercising**.

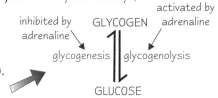

3) Adrenaline binds to **receptors** in the cell membrane of **liver cells**:
   - It **activates glycogenolysis** (the breakdown of glycogen to glucose).
   - It **inhibits glycogenesis** (the synthesis of glycogen from glucose).

4) It also **activates glucagon secretion** and **inhibits insulin secretion**, which increases glucose concentration.

5) Adrenaline gets the **body ready** for **action** by making **more glucose** available for **muscles** to respire.

6) Both **adrenaline** and **glucagon** can activate glycogenolysis **inside** a cell even though they bind to **receptors** on the **outside** of the cell. Here's **how** they do it:

- Adrenaline and glucagon **bind** to their specific receptors and **activate** an **enzyme** called **adenylate cyclase**.
- Activated adenylate cyclase converts **ATP** into a **chemical signal** called a '**second messenger**'.
- The second messenger is called **cyclic AMP (cAMP)**.
- cAMP **activates** a **cascade** (a chain of reactions) that break down glycogen into glucose (**glycogenolysis**).

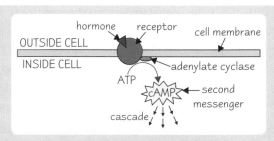

Diabetes Occurs when Blood Glucose Concentration is Not Controlled

Diabetes mellitus is a condition where **blood glucose** concentration **can't** be **controlled** properly. There are **two types**:

Type I

1) In **Type I** diabetes, the β **cells** in the islets of Langerhans **don't produce** any **insulin**.

2) After **eating**, the blood glucose level **rises** and **stays high** — this is called **hyperglycaemia** and can result in **death** if left untreated. The kidneys **can't reabsorb** all this glucose, so some of it's **excreted** in the urine.

3) It can be treated by regular **injections** of **insulin**. But this has to be **carefully controlled** because too much can produce a **dangerous drop** in blood glucose levels — this is called **hypoglycaemia**.

4) **Eating regularly** and **controlling simple carbohydrate intake** (sugars) helps to **avoid** a **sudden rise** in glucose.

Type II

1) **Type II** diabetes is usually acquired **later** in **life** than Type I, and it's often linked with **obesity**.

2) It occurs when the β cells **don't produce enough insulin** or when the body's **cells don't respond** properly to **insulin**. Cells don't respond properly because the insulin **receptors** on their membranes **don't work** properly, so the cells **don't** take up enough glucose. This means the **blood glucose concentration** is **higher** than normal.

3) It can be treated by **controlling simple carbohydrate intake** and **losing weight**. **Glucose-lowering tablets** can be taken if diet and weight loss can't control it.

Practice Questions

Q1 Give three functions of glucagon.

Q2 What is a second messenger?

Exam Questions

Q1 Describe and explain how hormones return blood glucose concentration to normal after a meal. [5 marks]

Q2 Explain why someone with diabetes can produce insulin but can't control their blood glucose concentration. [3 marks]

My α cells detect low glucose — urgent tea and biscuit break needed...

Aaaaargh there are so many stupidly complex names to learn and they all look and sound exactly the same to me.
You can't even get away with sneakily misspelling them all in your exam — like writing 'glycusogen' or 'glucogenesisolysis'.
Nope, examiners have been around for centuries, so I'm afraid old tricks like that just won't work on them. Grrrrrrr.

Control of the Menstrual Cycle

Sorry lads — these two pages are pretty much devoted to the inner workings of the ladies.
Now they're bound to start going on about how easy you have it in comparison. Got a point though...

The Human *Menstrual Cycle* is *Controlled* by *Hormones*

You don't need to learn this diagram (phew), but it shows you what's going on.

1) The human **menstrual cycle** (also called the **oestrous cycle**) lasts about **28 days**.

2) The menstrual cycle involves:

   - A **follicle** (an egg and its surrounding protective cells) **developing** in the **ovary**.
   - Ovulation — an **egg** being **released**.
   - The **uterus lining** becoming **thicker** so that a fertilised egg can **implant**.
   - A structure called a **corpus luteum** developing from the **remains** of the **follicle**.

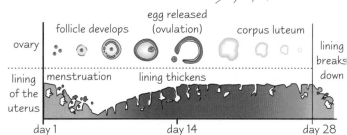

3) If there's **no fertilisation**, the uterus lining **breaks down** and leaves the body through the **vagina**. This is known as **menstruation**, which marks the **end** of one cycle and the **start** of another.

4) The menstrual cycle's **controlled** by the action of **four hormones**:

   - **Follicle-stimulating hormone** (**FSH**) — does just what it says, it **stimulates** the **follicle** to develop.
   - **Luteinising hormone** (**LH**) — **stimulates ovulation** and **stimulates** the **corpus luteum** to develop.
   - **Oestrogen** — **stimulates** the **uterus lining** to **thicken**.
   - **Progesterone** — **maintains** the **thick uterus lining**, ready for implantation of an embryo.

5) **FSH** and **LH** are secreted by the **anterior pituitary gland**. **Oestrogen** and **progesterone** are secreted by the **ovaries**.

Hormone Concentrations Change During *Different Stages* of the *Cycle*

1 **High FSH** concentration in the blood

- FSH stimulates **follicle development**.
- The **follicle** releases **oestrogen**.
- FSH **stimulates** the **ovaries** to release **oestrogen**.

2 **Rising** concentration of **oestrogen**

- Oestrogen **stimulates** the **uterus lining** to **thicken**.
- Oestrogen **inhibits FSH** being released from the **pituitary** gland.

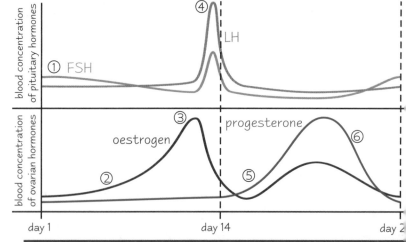

3 **Oestrogen** concentration **peaks**

High oestrogen concentration **stimulates** the pituitary gland to release **LH** and **FSH**.

4 **LH surge** (a rapid increase)

- **Ovulation** is stimulated by LH — the **follicle ruptures** and the **egg is released**.
- LH stimulates the **ruptured follicle** to turn into a **corpus luteum**.
- The **corpus luteum** releases **progesterone**.

5 **Rising** concentrations of **progesterone**

- Progesterone **inhibits FSH and LH release** from the **pituitary**.
- The **uterus lining** is **maintained** by progesterone.
- If **no embryo** implants, the **corpus luteum breaks down** and **stops releasing** progesterone.

6 **Falling** concentration of **progesterone**

- **FSH and LH** concentrations **increase** because they're **no longer inhibited** by progesterone.
- The uterus lining **isn't maintained** so it **breaks down** — **menstruation** happens and the **cycle starts again**.

Control of the Menstrual Cycle

Negative and Positive Feedback Mechanisms Control the Level of Hormones

The **different concentrations** of **hormones** in the blood during the menstrual cycle are **controlled** by **feedback loops**.

Negative feedback — example one

1) **FSH** stimulates the **ovary** to release **oestrogen**.

2) Oestrogen **inhibits** further release of **FSH**.

After FSH has stimulated follicle development, **negative feedback** keeps the **FSH** concentration **low**. This makes sure that **no more follicles develop**.

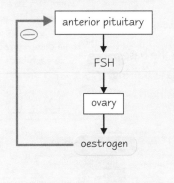

Negative feedback — example two

1) **LH** stimulates the **corpus luteum** to develop, which produces **progesterone**.

2) Progesterone **inhibits** further release of **LH**.

Negative feedback makes sure that **no more follicles develop** when the corpus luteum is developing. It also makes sure the uterus lining **isn't** maintained if **no embryo** implants.

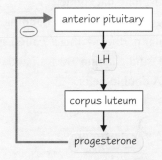

Positive feedback — example

1) **Oestrogen** stimulates the **anterior pituitary** to release **LH**.

2) **LH** stimulates the **ovary** to release **more oestrogen**.

3) Oestrogen **further stimulates** the **anterior pituitary** to release **LH**, and so on.

High oestrogen concentration triggers **positive feedback** to make **ovulation** happen.

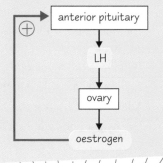

It's a bit weird — oestrogen usually inhibits FSH and LH release, but at really high levels it switches to stimulating their release.

Practice Questions

Q1 Name the hormones released by the anterior pituitary gland.

Q2 Which hormone stimulates the corpus luteum to develop?

Q3 What's the main role of oestrogen in the uterus?

Q4 What's the main role of progesterone in the uterus?

Q5 What happens to the uterus lining if no embryo implants?

Exam Questions

Q1 The human menstrual cycle is controlled by pituitary and ovarian hormones, which are present at different concentrations during the cycle.

a) Explain how negative feedback ensures only one main follicle develops. [5 marks]

b) Explain how positive feedback is involved in ovulation. [3 marks]

Q2 The contraceptive pill contains synthetic equivalents of the hormones oestrogen and progesterone. Suggest how taking the pill can prevent pregnancy. [3 marks]

Sometimes it's hard to be a woman...

What on earth... talk about women being hard to understand. Let's treat it like a good pair of shoes and take it in steps. Start by learning the four main hormones and what they do, then learn when concentrations are highest and why. Finally, get scribbling down those feedback loops. And when you know it all, it must be time for an end-of-section break.

DNA and RNA

*Deoxyribonucleic acid (a.k.a. **DNA**) is remarkably clever stuff — as you might remember from AS. Not only do you revisit it for A2, you get to go a step further and head into the mysterious world of RNA. Let the good times roll.*

DNA is Made of **Nucleotides** that Contain a **Sugar**, a **Phosphate** and a **Base**

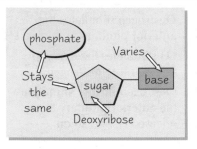

1) DNA is a **polynucleotide** — it's made up of lots of **nucleotides** joined together.
2) Each nucleotide is made from a **pentose sugar** (with 5 carbon atoms), a **phosphate** group and a **nitrogenous base**.
3) The **sugar** in DNA nucleotides is a **deoxyribose** sugar.
4) Each nucleotide has the **same sugar and phosphate**. The **base** on each nucleotide can **vary** though.
5) There are **four** possible bases — adenine (**A**), thymine (**T**), cytosine (**C**) and guanine (**G**).

You learnt a lot of this at AS but you need to know it for A2 as well.

Two Polynucleotide Strands **Join Together** to Form a **Double-Helix**

1) DNA nucleotides join together to form **polynucleotide strands**.
2) The nucleotides join up between the **phosphate** group of one nucleotide and the **sugar** of another, creating a **sugar-phosphate backbone**.
3) **Two** DNA polynucleotide strands join together by **hydrogen bonding** between the bases.
4) Each base can only join with one particular partner — this is called **specific base pairing**.
5) **Adenine** always pairs with **thymine** (**A - T**) and **guanine** always pairs with **cytosine** (**G - C**).
6) The two strands **wind up** to form the **DNA double-helix**.

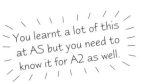

When two strands have bases that pair up the strands are said to be complementary to each other:

```
A T C G G
| | | | |
T A G C C
```

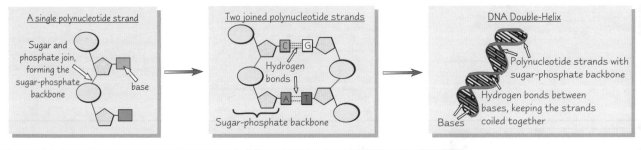

A single polynucleotide strand

Sugar and phosphate join, forming the sugar-phosphate backbone

base

Two joined polynucleotide strands

Hydrogen bonds

Sugar-phosphate backbone

DNA Double-Helix

Polynucleotide strands with sugar-phosphate backbone

Hydrogen bonds between bases, keeping the strands coiled together

Bases

DNA Contains **Genes** Which are **Instructions** for **Proteins**

1) Genes are **sections of DNA**. They're found on **chromosomes**.
2) Genes **code** for **proteins** (polypeptides) — they contain the **instructions** to make them.
3) Proteins are made from **amino acids**. Different proteins have a **different number** and **order** of amino acids.
4) It's the **order** of **nucleotide bases** in a gene that determines the **order of amino acids** in a particular **protein**.
5) Each amino acid is coded for by a sequence of **three bases** (called a **triplet**) in a gene. A DNA triplet is also called a **base triplet** or a **codon**.
6) **Different sequences** of bases code for different amino acids. This is the **genetic code** — see page 80 for more.
7) So the **sequence of bases** in a section of DNA is a **template** that's used to make a **protein** during **protein synthesis**.

Polypeptide is just another word for a protein.

Bases on DNA

G T C T G A

DNA triplet = one amino acid

DNA is **Copied** into **RNA** for **Protein Synthesis**

1) DNA molecules are found in the **nucleus** of the cell, but the organelles for protein synthesis (**ribosomes**) are found in the **cytoplasm**.
2) DNA is too large to move out of the nucleus, so a section is **copied** into **RNA**. This process is called **transcription** (see page 78).
3) The RNA **leaves** the nucleus and joins with a **ribosome** in the cytoplasm, where it can be used to synthesise a **protein**. This process is called **translation** (see page 79).

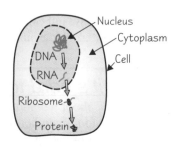

Nucleus

Cytoplasm

Cell

DNA

RNA

Ribosome

Protein

DNA and RNA

RNA is Very Similar to DNA

Like DNA, **RNA** (**r**ibo**n**ucleic **a**cid) is made of **nucleotides** that contain one of **four different bases**. The nucleotides also form a **polynucleotide strand** with a sugar-phosphate backbone. But RNA **differs** from DNA in **three** main ways:

1) The **sugar** in RNA nucleotides is a **ribose sugar** (not deoxyribose).

2) The nucleotides form a **single polynucleotide strand** (not a double one).

3) **Uracil** (**U**) replaces thymine as a base. Uracil **always pairs** with **adenine** during protein synthesis.

You need to know about **two different types** of RNA — **messenger RNA** (**mRNA**) and **transfer RNA** (**tRNA**).

Messenger RNA (mRNA)

mRNA is a **single polynucleotide strand**. In mRNA, groups of three adjacent bases are usually called **codons** (they're sometimes called **triplets** or **base triplets**). mRNA is made in the **nucleus** during **transcription**. It **carries the genetic code** from the DNA in the **nucleus** to the **cytoplasm**, where it's used to make a **protein** during **translation**.

Base
A
C — codon
U
Phosphate
Ribose sugar

Transfer RNA (tRNA)

tRNA is a **single polynucleotide strand** that's folded into a **clover shape**. **Hydrogen bonds** between **specific base pairs** hold the molecule in this shape. Every tRNA molecule has a **specific sequence** of **three bases** at one end called an **anticodon**. They also have an **amino acid binding site** at the other end. tRNA is found in the **cytoplasm** where it's involved in **translation**. It **carries** the amino acids that are used to make **proteins** to the **ribosomes**.

A C C — Amino acid binding site
Anticodon
G A U
Hydrogen bonds between base pairs

You need to be able to Compare DNA, mRNA and tRNA

You need to know the **structure** and **composition** of **DNA**, **mRNA** and **tRNA** really well — you could be asked to **compare** them in your exam. The table below outlines the **main differences** between them:

	DNA	mRNA	tRNA
Shape	Double-stranded — twisted into a double-helix and held together by hydrogen bonds	Single-stranded	Single-stranded — folded into a clover shape and held together by hydrogen bonds
Sugar	Deoxyribose sugar	Ribose sugar	Ribose sugar
Bases	A, T, C, G	A, U, C, G	A, U, C, G
Other features	Three adjacent bases are called a triplet (sometimes a base triplet or codon)	Three adjacent bases are called a codon (sometimes a triplet or base triplet)	Each tRNA molecule has a specific sequence of three bases called an anticodon and an amino acid binding site

tRNA growing in its natural environment.

Practice Questions

Q1 What are the nucleotides in DNA made of?

Q2 Name the bases found in DNA.

Q3 Describe the shape of a tRNA molecule.

Exam Question

Q1 a) Describe the differences in the composition of DNA and RNA molecules. [2 marks]

b) Name each of the following molecules:

i) A single-stranded molecule that contains ribose sugar and has an amino acid binding site. [1 mark]

ii) A double-stranded molecule that contains deoxyribose sugars and the base thymine. [1 mark]

iii) A single-stranded molecule that contains the base uracil and has an anticodon. [1 mark]

Genes, genes are good for your heart, the more you eat, the more you...

An easy way to remember where mRNA and tRNA come into the whole protein synthesis game is to look at the first letters. mRNA is a messenger — it carries the code from DNA to a ribosome. tRNA transfers amino acids. Easy as that.

Protein Synthesis

OK, so you know all about DNA and RNA — what they're made of, what they look like and the different types. Now you find out what they actually do. It gets kind of complicated but bear with it — it's impressive stuff.

First Stage of Protein Synthesis — Transcription

During transcription an **mRNA copy** of a gene (a section of DNA) is made in the **nucleus**:

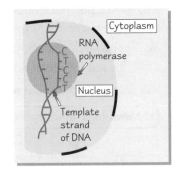

1) Transcription starts when **RNA polymerase** (an **enzyme**) **attaches** to the **DNA** double-helix at the **beginning** of a **gene**.

2) The **hydrogen bonds** between the two DNA strands in the gene **break**, **separating** the strands, and the DNA molecule **uncoils** at that point.

3) One of the strands is then used as a **template** to make an **mRNA copy**.

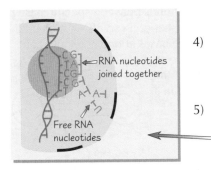

4) The RNA polymerase lines up free **RNA nucleotides** alongside the template strand. **Specific base pairing** means that the mRNA strand ends up being a **complementary copy** of the DNA template strand (except the base **T** is replaced by **U** in **RNA**).

5) Once the RNA nucleotides have **paired up** with their **specific bases** on the DNA strand they're **joined together**, forming an **mRNA** molecule.

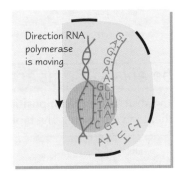

6) The RNA polymerase moves **along** the DNA, separating the strands and **assembling** the mRNA strand.

7) The **hydrogen bonds** between the uncoiled strands of DNA **re-form** once the RNA polymerase has passed by and the strands **coil back into a double-helix**.

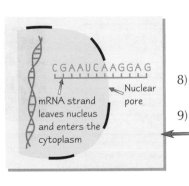

8) When RNA polymerase reaches a particular sequence of DNA called a **stop signal**, it stops making mRNA and **detaches** from the DNA.

9) The **mRNA** moves **out** of the **nucleus** through a nuclear pore and attaches to a **ribosome** in the cytoplasm, where the next stage of protein synthesis takes place (see next page).

mRNA is Edited in Eukaryotic Cells

1) Genes in **eukaryotic DNA** contain sections that **don't code** for amino acids.

2) These sections of DNA are called **introns**. All the bits that **do** code for amino acids are called **exons**.

3) During transcription the introns and exons are both **copied** into mRNA. mRNA strands containing introns and exons are called **pre-mRNA**.

4) Introns are **removed** from pre-mRNA strands by a process called **splicing** — introns are removed and exons joined forming **mRNA** strands. This takes place in the **nucleus**.

5) The mRNA then **leaves** the nucleus for the next stage of protein synthesis (**translation**).

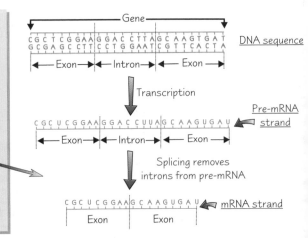

Protein Synthesis

Second Stage of Protein Synthesis — Translation

Translation occurs at the **ribosomes** in the **cytoplasm**. During **translation**, **amino acids** are **joined together** to make a **polypeptide chain** (protein), following the sequence of **codons** (triplets) carried by the mRNA.

1) The **mRNA attaches** itself to a **ribosome** and **transfer RNA (tRNA)** molecules **carry amino acids** to the ribosome.

2) A tRNA molecule, with an **anticodon** that's **complementary** to the **first codon** on the mRNA, attaches itself to the mRNA by **specific base pairing**. ⟹

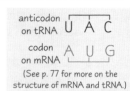

anticodon on tRNA U A C

codon on mRNA A U G

(See p. 77 for more on the structure of mRNA and tRNA.)

3) A second tRNA molecule attaches itself to the **next codon** on the mRNA in the **same way**.

4) The two amino acids attached to the tRNA molecules are **joined** by a **peptide bond**. The first tRNA molecule **moves away**, leaving its amino acid behind.

5) A third tRNA molecule binds to the **next codon** on the mRNA. Its amino acid **binds** to the first two and the second tRNA molecule **moves away**.

6) This process continues, producing a chain of linked amino acids (a **polypeptide chain**), until there's a **stop signal** on the mRNA molecule.

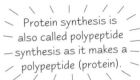

Protein synthesis is also called polypeptide synthesis as it makes a polypeptide (protein).

7) The polypeptide chain (**protein**) **moves away** from the ribosome and translation is complete.

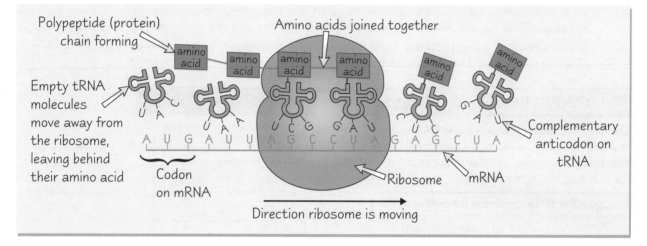

Polypeptide (protein) chain forming — amino acid

Amino acids joined together

Empty tRNA molecules move away from the ribosome, leaving behind their amino acid

Codon on mRNA

Ribosome

mRNA

Complementary anticodon on tRNA

Direction ribosome is moving

Practice Questions

Q1 What are the two stages of protein synthesis called?

Q2 Where does the first stage of protein synthesis take place?

Q3 When does RNA polymerase stop making mRNA?

Q4 What is an exon?

Q5 Where does the second stage of protein synthesis take place?

Q6 What is a polypeptide chain (protein) made up of?

Exam Questions

Q1 A drug that inhibits cell growth is found to be able to bind to DNA, preventing RNA polymerase from binding. Explain how this drug will affect protein synthesis. [2 marks]

Q2 A polypeptide chain (protein) from a eukaryotic cell is 10 amino acids long.
   a) Predict how long the mRNA for this protein would be in nucleotides. Explain your answer. [2 marks]
   b) Would you expect the number of nucleotides in the gene (DNA sequence) for this protein to be greater or fewer than your answer for part a)? Explain your answer. [3 marks]

The only translation I'm interested in is a translation of this page into English...

So you start off with DNA, lots of cleverness happens and bingo... you've got a protein. Only problem is you need to know the cleverness bit in quite a lot of detail. So scribble it down, recite it to yourself, explain it to your best mate or do whatever else helps you remember the joys of protein synthesis. And then think how clever you must be to know it all.

The Genetic Code and Nucleic Acids

The genetic code is exactly as it sounds — a code found in your genes that tells your body how to make proteins. It can be interpreted, just like any other code, which is exactly what you might have to do in your exam. So get cracking.

The Genetic Code is **Non-Overlapping**, **Degenerate** and **Universal**

1)   The genetic code is the **sequence of base triplets** (codons) in **mRNA** which **code** for specific **amino acids**.

2)   In the genetic code, each base triplet is **read** in sequence, **separate** from the triplet **before** it and **after** it. Base triplets **don't share** their **bases** — the code is **non-overlapping**.

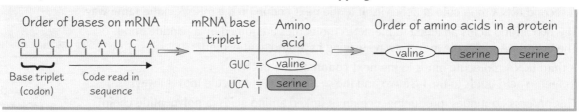

3)   The genetic code is also **degenerate** — there are **more** possible combinations of **triplets** than there are amino acids (20 amino acids but 64 possible triplets). This means that some **amino acids** are coded for by **more than one** base triplet, e.g. tyrosine can be coded for by UAU or UAC.

4)   Some triplets are used to tell the cell when to **start** and **stop** production of the protein — these are called **start** and **stop** signals (or **codons**). They're found at the **beginning** and **end** of the mRNA. E.g. UAG is a stop signal.

5)   The genetic code is also **universal** — the **same** specific base triplets code for the **same** amino acids in **all living things**. E.g. UAU codes for tyrosine in all organisms.

You need to be able to **Interpret Data** about **Nucleic Acids**

The table on the right shows the **mRNA codons** (triplets) for some amino acids. You might have to **interpret** information like this in the exam. For example, using the table, you could be asked to...

mRNA codon	Amino Acid
UCU	Serine
CUA	Leucine
UAU	Tyrosine
GUG	Valine
GCA	Alanine
CGC	Arginine

When interpreting data on nucleic acids remember that DNA contains T and RNA contains U.

...give the DNA sequence for amino acids

The mRNA codons for the amino acids are given in the table. Because **mRNA** is a **complementary copy** of the **DNA** template, the DNA sequence for each amino acid is made up of bases that would **pair** with the mRNA sequence:

mRNA codon	Amino Acid	DNA sequence (of template strand)
UCU	Serine	AGA
CUA	Leucine	GAT
UAU	Tyrosine	ATA
GUG	Valine	CAC
GCA	Alanine	CGT
CGC	Arginine	GCG

You could also be asked to work out the amino acids from a given DNA sequence and a table.

...give the tRNA anticodons from mRNA codons

tRNA anticodons are **complementary copies** of **mRNA codons**, so you can work out the tRNA anticodon from the mRNA codon:

mRNA codon	tRNA anticodon
UCU	AGA
CUA	GAU
UAU	AUA
GUG	CAC
GCA	CGU
CGC	GCG

You might be asked to name the amino acid coded for by a tRNA anticodon using a table like the one above.

...write the amino acid sequence for a section of mRNA

To **work out** the sequence of **amino acids** from some mRNA, you need to break the genetic code into **codons** and then use the information in the table to work out what **amino acid** they code for.

You might have to work out the sequence of some mRNA from a sequence of amino acids and a table.

Example

mRNA: CUAGUGCGCUAUUCU

Codons: CUA   GUG   CGC   UAU   UCU

Amino acids: Leucine   Valine   Arginine   Tyrosine   Serine

The Genetic Code and Nucleic Acids

In the exam you might have to **interpret data** from experiments done to **investigate nucleic acids** and their **role** in **protein synthesis**. Here's an example (you **don't** need to **learn** it):

Investigating the effect of new drugs on nucleic acids

1) To investigate **how** two new drugs affect **nucleic acids** and their **role** in protein synthesis, **bacteria** were **grown** in **normal conditions** for a few generations, then moved to media containing the drugs.

2) After a short period of time, the **concentration** of **protein** and **complete strands** of **mRNA** in the bacteria were analysed. The results are shown in the **bar graph**.

3) Both mRNA **and** protein concentration were **lower** in the presence of **drug 1** compared to the **no-drug control**. This suggests that drug 1 **affects the production** of **full length mRNA**, so there's no mRNA for protein synthesis during **translation**.

4) **mRNA production** in the presence of **drug 2** was **unaffected**, but **less protein** was produced — **3** mg cm^{-3} compared to **8** mg cm^{-3}. This suggests that drug 2 **interferes** with **translation**. **mRNA was produced**, but **less protein** was **translated** from it.

5) **Further tests** to establish the **nature** of the two drugs were carried out.

6) **Drug 1** was found to be a **ribonuclease** (an enzyme that **digests RNA**). This could **explain** the results of the first experiment — **any strands** of **mRNA** produced by the cell would be **digested** by drug 1, so **couldn't be used** in **translation** to make proteins.

7) **Drug 2** was found to be a **single-stranded**, **clover-shaped** molecule capable of binding to the **ribosome**. Again, this helps to **explain** the **results** from the first experiment — drug 2 could work by **binding** to the ribosome, **blocking tRNAs** from binding to it and so **preventing translation**.

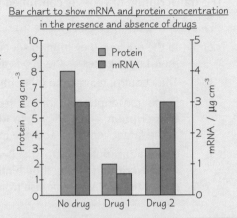

Bar chart to show mRNA and protein concentration in the presence and absence of drugs

Transcription and translation are on pages 78-79.

Practice Questions

Q1 What is the genetic code?

Q2 Why is the genetic code described as degenerate?

Q3 Why is the genetic code described as universal?

mRNA codon	amino acid
UGU	Cysteine
CGC	Arginine
GGG	Glycine
GUG	Valine
GCA	Alanine
UUG	Leucine
UUU	Phenylalanine

Exam Questions

Q1 An artificial mRNA was synthesized and used in an experiment to form a protein (polypeptide). The mRNA sequence was: UUGUGUGGGUUUGCAGCA and the protein produced was: Leucine–Cysteine–Glycine–Phenylalanine–Alanine–Alanine. Use the table above to help you answer the following questions.

a) Explain how the result suggests that the genetic code is based on triplets of nucleotides in mRNA. [2 marks]

b) Explain how the result suggests that the genetic code is non-overlapping. [2 marks]

Q2 The table shows the mRNA codons for some amino acids. Show your working for the following questions.

a) Give the amino acid sequence for the mRNA sequence GUGUGUCGCGCA. [2 marks]

b) Give the mRNA sequence for the amino acid sequence arginine, alanine, leucine, phenylalanine. [2 marks]

c) Give the DNA template strand sequence that codes for the amino acid sequence valine, arginine, alanine. [3 marks]

Hurrah — a page with slightly fewer confusing terms and a lot less to remember. The key to the genetic code is to be able to interpret it, so if you know how DNA, mRNA and tRNA work together to make a protein you should be able to handle any data they can throw at you. Now repeat after me, C pairs with G, A pairs with T. Unless it's RNA — then it's U.

Regulation of Transcription and Translation

Oh yes, you read that right — it's back to the incredibly important and immensely clever transcription and translation.

Transcription Factors Control the Transcription of Target Genes

All the **cells** in an organism carry the **same genes** (DNA) but the **structure** and **function** of different cells **varies**. This is because **not all** the **genes** in a cell are **expressed** (transcribed and used to make a protein). Because **different genes** are expressed, **different proteins** are made and these proteins modify the cell — they determine the **cell structure** and control **cell processes** (including the expression of more genes, which produce more proteins).

The **transcription** of genes is **controlled** by protein molecules called **transcription factors**:

1) Transcription factors **move** from the **cytoplasm** to the **nucleus**.

2) In the nucleus they **bind** to **specific DNA sites** near the start of their **target genes** — the genes they **control** the expression of.

3) They control expression by controlling the **rate** of transcription.

4) Some transcription factors, called **activators**, **increase** the **rate of transcription** — e.g. they help **RNA polymerase bind** to the start of the target gene and **activate** transcription.

5) Other transcription factors, called **repressors**, **decrease** the **rate of transcription** — e.g. they **bind** to the start of the target gene, **preventing RNA polymerase** from **binding**, **stopping** transcription.

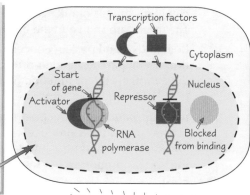

Transcription and translation are covered on pages 78-79.

Oestrogen Affects the Transcription of Target Genes

The **expression** of genes can also be **affected** by **other molecules** in the cell, e.g. **oestrogen**:

1) Oestrogen is a **hormone** that can affect transcription by **binding** to a **transcription factor** called an **oestrogen receptor**, forming an **oestrogen-oestrogen receptor complex**.

2) The complex moves from the **cytoplasm** into the **nucleus** where it **binds** to **specific DNA sites** near the **start** of the **target gene**.

3) The complex can **either** act as an **activator**, e.g. **helping** RNA polymerase, or as a **repressor**, e.g. **blocking** RNA polymerase.

4) Whether the complex **acts** as a repressor **or** activator **depends** on the **type of cell** and the target gene.

5) So, the **level of oestrogen** in a particular cell affects the **rate of transcription** of target genes.

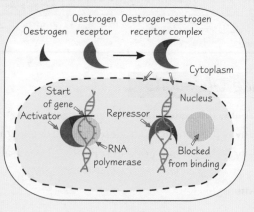

siRNA Interferes with Gene Expression

Gene expression is also affected by a **type of RNA** called **small interfering RNA (siRNA)**:

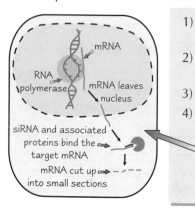

1) siRNA molecules are **short**, **double-stranded RNA** molecules that can **interfere** with the **expression** of specific genes.

2) Their bases are **complimentary** to **specific sections** of a **target gene** and the mRNA that's formed from it.

3) siRNA can **interfere** with both the **transcription** and **translation** of genes.

4) siRNA affects **translation** through a mechanism called **RNA interference**:

• In the **cytoplasm**, siRNA and associated proteins **bind** to the **target mRNA**.

• The proteins **cut up** the mRNA into sections so it can no longer be translated.

• So, the siRNA **prevents the expression** of the specific gene as its protein can no longer be made during **translation**.

Regulation of Transcription and Translation

You need to be able to Interpret Experimental Data on Gene Expression

You could get a question in the exam where you have to **interpret data** about gene expression. It could be on **anything** you've learnt on the **previous page**, e.g. transcription factors, oestrogen or siRNAs. Below is an example of a **gene expression system** in bacteria and an experiment that **investigates** how it works. You **don't** need to **learn** the information, just **understand** what the results of the experiment tell you about how the expression of the gene is **controlled**.

Transcribing — I can do it with my eyes closed.

The *lac* repressor:

1) *E. coli* is a **bacterium** that respires **glucose**, but it can use **lactose** if glucose **isn't available**.

2) If lactose is present, *E. coli* makes an **enzyme** (β-galactosidase) to **digest** it. But if there's **no** lactose, it doesn't **waste energy** making an enzyme it **doesn't need**. The enzyme's **gene** is **only expressed** when lactose is **present**.

3) The production of the enzyme is **controlled** by a **transcription factor** — the *lac* **repressor**.

4) When there's **no** lactose, the *lac* repressor **binds** to the DNA at the start of the gene, **stopping transcription**.

5) When lactose **is present** it **binds** to the *lac* repressor, **stopping** it binding to the DNA, so the gene **is transcribed**.

Experiment:

1) Different *E. coli* mutants were isolated and grown in **different media**, e.g. with lactose or glucose.

2) The mutants have **mutations** (**changes** in their **DNA bases**, see next page) that mean they **act differently** from normal *E. coli*, e.g. they **produce** β-galactosidase when grown with glucose.

3) To **detect** whether active (working) β-galactosidase was produced, a **chemical** that turns **yellow** in the presence of active β-galactosidase was **added** to the medium.

Medium	Mutant	mRNA	Colour
Glucose	Normal	No	No yellow
Lactose	Normal	Yes	Yellow
Glucose	Mutant 1	Yes	Yellow
Lactose	Mutant 1	Yes	Yellow
Glucose	Mutant 2	No	No yellow
Lactose	Mutant 2	Yes	No yellow

4) The production of **mRNA** that **codes** for β-galactosidase was also measured. The results are shown in the **table**.

5) In **mutant 1**, mRNA and active β-galactosidase **were produced** even when they were grown with **only glucose** — the gene is **always** being expressed.

6) This suggests that mutant 1 has a **faulty** *lac* **repressor**, e.g. in the **absence** of lactose the repressor **isn't able** to bind DNA, so transcription **can** occur and mRNA and active β-galactosidase **are produced**.

7) In **mutant 2**, mRNA is produced but active **β-galactosidase isn't** when **lactose** is present — the **gene** is being **transcribed** but it **isn't** producing active β-galactosidase.

8) This suggests mutant 2 is producing **faulty β-galactosidase**, e.g. because a **mutation** has affected its active site.

Practice Questions

Q1 Name two types of transcription factor.

Q2 Name the transcription factor that oestrogen can bind to.

Q3 What does siRNA stand for?

Tube	Medium	Bacteria	Full length mRNA	Protein
1	+ Oestrogen	Normal	Yes	Active
2	− Oestrogen	Normal	No	No
3	+ Oestrogen	Mutant	No	No
4	− Oestrogen	Mutant	No	No

Exam Question

Q1 An experiment was carried out to investigate gene expression of the Chi protein in genetically engineered bacteria. A mutant bacterium was isolated and analysed to look for mRNA coding for Chi, and active Chi protein production. The results are shown in the table above.

a) What do the results of tubes 1 and 2 suggest about the control of gene expression? Explain your answer. [4 marks]

b) What do the results of tubes 3 and 4 suggest could be wrong with the mutant? Explain your answer. [3 marks]

c) If an siRNA complimentary to the Chi gene was added to tube 1, what would you expect the results to be? Explain your answer. [3 marks]

Transcription Factor — not quite as eXciting as that other factor programme...

If it was a competition, oestrogen would totally win — it's very jazzy and awfully controlling. Flexible too — sometimes it helps to activate and other times it helps to repress. Although I'm not sure it can hold a note or wiggle in time to music.

Mutations, Genetic Disorders and Cancer

Mutations — as featured in numerous superhero movies. Well, I'm sorry to be the one to break it to you, but they don't usually give you special powers or superhuman strength — in fact, they can cause a lot of problems.

Mutations are Changes to the Base Sequence of DNA

Any change to the **base sequence** of DNA is called a **mutation**:

1) They can be caused by **errors** during **DNA replication**.

2) They can also be caused by **mutagenic agents** (see below).

Errors can also be caused by insertion, duplication and inversion of bases.

3) The **types** of errors that can occur include:

- **Substitution** — one base is substituted with another, e.g. AT**G**CCT becomes AT**T**CCT (G is **swapped** for T).
- **Deletion** — one base is deleted, e.g. AT**G**CCT becomes ATCCT (G is **deleted**).

4) The **order** of **DNA bases** in a gene determines the **order of amino acids** in a particular **protein** (see p. 76). If a mutation occurs in a gene, the **sequence** of **amino acids** that it codes for (and the protein formed) could be **altered**.

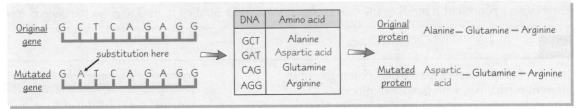

Not All Mutations Affect the Order of Amino Acids

The **degenerate nature** of the genetic code (see page 80) means that some amino acids are coded for by **more than one DNA triplet** (e.g. tyrosine can be coded for by TAT or TAC in DNA). This means that **not all** substitution mutations will result in a change to the amino acid sequence of the protein — some substitutions will still **code** for the **same amino acid**. For example:

DNA	Amino acid
TAT	Tyrosine
TAC	Tyrosine
AGT	Serine
CTT	Leucine
GTC	Valine

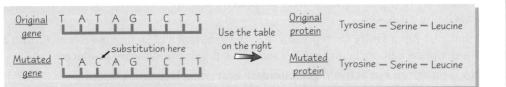

Substitution mutations **won't always** lead to changes in the amino acid sequence, but **deletions will** — the deletion of a base will change the **number** of bases present, which will cause a **shift** in all the base triplets after it:

Mutagenic Agents Increase the Rate of Mutation

Mutations occur **spontaneously**, e.g. when DNA is **misread** during **replication**. But some things can cause an **increase** in the **rate** of mutations — these are called **mutagenic agents**. **Ultraviolet radiation, ionising radiation**, some **chemicals** and some **viruses** are examples of mutagenic agents. They can increase the rate of mutations by:

1) **Acting as a base** — chemicals called **base analogs** can **substitute** for a base during DNA replication, **changing** the **base sequence** in the new DNA. E.g. **5-bromouracil** is a base analog that can substitute for **thymine**. It can pair with **guanine** (**instead** of **adenine**), causing a **substitution mutation** in the new DNA.

2) **Altering bases** — some chemicals can **delete** or **alter** bases. E.g. **alkylating agents** can add an alkyl group to **guanine**, which **changes** the **structure** so that it pairs with **thymine** (**instead** of **cytosine**).

3) **Changing the structure of DNA** — some types of **radiation** can change the structure of DNA, which causes **problems** during DNA replication. E.g. **UV radiation** can cause adjacent **thymine** bases to **pair up** together.

Mutations, Genetic Disorders and Cancer

Genetic Disorders and Cancer are Caused By Mutations

Hereditary Mutations Cause Genetic Disorders and Some Cancers

Some mutations can cause **genetic disorders** — inherited disorders caused by **abnormal genes** or **chromosomes**, e.g. cystic fibrosis. Some mutations can **increase** the **likelihood** of developing certain **cancers**, e.g. mutations of the gene **BRCA1** can increase the chances of developing **breast cancer**. If a **gamete** (sex cell) containing a mutation for a genetic disorder or certain cancer is **fertilised**, the mutation will be present in the new **fetus** formed — these are called **hereditary mutations** because they are passed on to the offspring.

Acquired Mutations Can Cause Cancer

1) Mutations that occur in individual cells **after** fertilisation (e.g. in adulthood) are called **acquired mutations**.

2) If these mutations occur in the **genes** that **control** the rate of **cell division**, it can cause **uncontrolled cell division**.

3) If a cell divides uncontrollably the result is a **tumour** — a mass of abnormal cells. Tumours that **invade** and **destroy surrounding tissue** are called **cancers**.

4) There are **two types** of **gene** that control cell division — **tumour suppressor genes** and **proto-oncogenes**. Mutations in these genes can cause cancer:

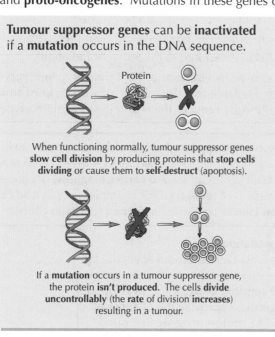

Tumour suppressor genes can be **inactivated** if a **mutation** occurs in the DNA sequence.

When functioning normally, tumour suppressor genes **slow cell division** by producing proteins that **stop cells dividing** or cause them to **self-destruct** (apoptosis).

If a **mutation** occurs in a tumour suppressor gene, the protein **isn't produced**. The cells **divide uncontrollably** (the **rate** of division **increases**) resulting in a tumour.

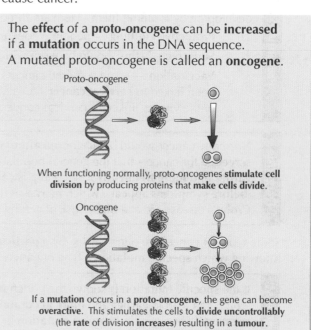

The **effect** of a **proto-oncogene** can be **increased** if a **mutation** occurs in the DNA sequence. A mutated proto-oncogene is called an **oncogene**.

When functioning normally, proto-oncogenes **stimulate cell division** by producing proteins that **make cells divide**.

If a **mutation** occurs in a **proto-oncogene**, the gene can become **overactive**. This stimulates the cells to **divide uncontrollably** (the **rate** of division **increases**) resulting in a **tumour**.

Practice Questions

Q1 What is a substitution mutation?

Q2 What are mutagenic agents?

Q3 What is a genetic disorder?

Before exposure	A	G	T	T	A	T	C	A	G	G	C	T

After exposure	A	G	G	T	A	T	G	A	G	G	C	C

DNA	Amino acids	DNA	Amino acids
AGT	Serine	GAG	Glutamic acid
AGG	Arginine	GCT	Alanine
TAT	Tyrosine	GCC	Alanine
CAG	Glutamine		

Exam Question

Q1 The order of bases in a liver cell's proto-oncogene before and after exposure to a mutagenic agent is shown above.

a) Underline any mutation(s) that have occurred. [1 mark]

b) Use the table to explain the changes that the mutations would cause to the sequence of amino acids. [5 marks]

c) Would you describe these mutations as acquired or inherited? Explain your answer. [2 marks]

d) Explain how the mutation(s) may lead to cancer. [3 marks]

Just hope your brain doesn't have a deletion mutation during the exam...

Right, there's plenty to learn on these pages and some of it's a bit complicated, so you know the drill. Don't read it all through at once — take the sections one by one and get all the facts straight. There could be nothing more fun...

Diagnosing and Treating Cancer and Genetic Disorders

Before you start this page, make sure you've got one thing straight in your head from the previous page — make sure you understand the difference between acquired and hereditary disorders.

Knowing the Mutation is Useful for the Diagnosis and Treatment of Disorders

1) **Cancer** and most **genetic disorders** are caused by **mutations** (see previous page).

2) Knowing whether a disorder is caused by an **acquired** or **inherited mutation** affects the **prevention** and **diagnosis** of the disorder.

3) **Identifying** the **specific mutation** that causes a disorder in an individual affects the prevention, diagnosis and treatment.

See page 102 for how to screen for mutations.

4) Here are some **examples** for each type of **disorder**:

1) Cancer — Caused by Acquired Mutations

Cancer associated with hereditary mutations is covered on the next page.

Acquired mutations can **occur spontaneously** or be **caused by** exposure to **mutagenic agents** (see page 84). Knowing that cancer can be caused by **acquired mutations** affects the prevention and diagnosis:

> **Prevention**
>
> If you know that **acquired mutations** are caused by **mutagenic agents** you can try to prevent cancer developing by **avoiding them**. Here are three ways mutagenic agents can be avoided:
>
> 1) **Protective clothing** — people who **work** with mutagenic agents should wear protective clothing.
>
> 2) **Sunscreen** — this should be worn when the skin is exposed to the **Sun** (UV radiation).
>
> 3) **Vaccination** — some acquired cancers are caused by **viruses**, e.g. **HPV** (human papillomavirus) has been linked to **cervical cancer**. A vaccine is available that should protect women from **around 80%** of the viruses linked to cervical cancer. This greatly **reduces the risk** of developing this type of cancer.

> **Diagnosis**
>
> Normally cancer would be diagnosed **after symptoms** had **appeared**. **High-risk individuals** can be **screened** for cancers that the general population aren't normally screened for. Or they can be **screened earlier** and **more frequently** if screening is carried out. This can lead to **earlier diagnosis** of cancer (**before symptoms appear**), which **increases** the chances of **recovery**. For example, people who have **Crohn's disease** are at a higher risk of getting **colon cancer** and so are **screened** for colon cancer.

Some **types** of cancer are often caused by a **particular mutation**.
Knowing which specific mutation a type of cancer is usually caused by can affect diagnosis:

> **Diagnosis**
>
> If the **specific mutation** is known then often **more sensitive tests** can be **developed**, which can lead to **earlier** and **more accurate diagnosis**, improving the **chances of recovery**. For example, there's a **mutation** in the **RAS proto-oncogene** in around **half** of all **bowel cancers**. Bowel cancer can be **detected early** by looking for RAS mutations in the DNA of **bowel cells**.

Individuals diagnosed with **cancer** can also have the **DNA** from the **cancerous cells analysed** to see which mutation has caused it. Knowing **which specific mutation** the cancer is caused by affects treatment:

> **Treatment**
>
> 1) The **treatment** can be **different** for different mutations. For example, **breast cancer** caused by mutation of the **HER2 proto-oncogene** can be treated with a drug called **Herceptin®**. This drug binds **specifically** to the altered HER2 protein receptor and **suppresses cell division and tumour growth**. Breast cancer caused by other mutations is not treated with this drug as it doesn't work.
>
> 2) The **aggressiveness** of the **treatment** can **differ** depending on the mutation. Different mutations produce **different types** of cancer, which affects the treatment. For example, if the mutation is known to cause an **aggressive** (**fast-growing**) cancer it may be treated with **higher doses** of **radiotherapy** or by **removing larger areas** of the tumour and surrounding tissue during **surgery**.
>
> 3) If the specific mutation is known, **gene therapy** (see page 104) may be able to treat it. For example, if you know it's caused by **inactivated tumour suppressor genes** (see previous page), gene therapy could be used to provide **working versions** of the genes.

Diagnosing and Treating Cancer and Genetic Disorders

(2) *Cancer* — Caused by **Hereditary** Mutations

Cancer caused by **hereditary mutations** usually results in a **family history** of a certain type of cancer. If an individual has a family history of cancer, things can be done to **prevent** it **developing** and **diagnose it earlier** if it does:

Prevention
Most cancers are caused by mutations in **multiple genes**. So people with a family history should **avoid gaining extra acquired mutations** by **avoiding mutagenic agents**, e.g. those with a family history of **lung cancer shouldn't smoke**.

Diagnosis
Screening, or **increased** and **earlier screening**, if there's a **family history** can lead to **early detection** (i.e. before symptoms appear) and **increased** chances of recovery. E.g. **more frequent breast examinations** if there's a family history of **breast cancer**.

Individuals with a **family history** of cancer can have their **DNA analysed** to see if they carry the **specific mutation**. Knowing **which specific mutation** the cancer is caused by affects prevention, diagnosis and treatment:

Prevention
If the mutation causes a very high risk of cancer **preventative surgery** may be carried out — removing the **organ** the cancer is likely to affect **before cancer develops**. E.g. women with a mutation in BRCA1 sometimes choose to have a **mastectomy** (removal of one or both breasts) to **prevent breast cancer** from developing.

Diagnosis
Screening, or **increased** and **early screening** of those with a **hereditary mutation** can lead to **early detection** and **increased** chances of recovery. E.g. **frequent colonoscopies** for those with a mutated **APC gene** to diagnose hereditary **colon cancer earlier**.

Treatment
Treatment is **similar** to treating cancer caused by acquired mutations (see previous page). E.g. the treatment depends on the **particular mutation**. But cancer caused by hereditary mutations is often **diagnosed earlier**, which can **change** the treatment used.

(3) *Genetic Disorders* — Caused by **Hereditary** Mutations

Diagnosis
If a person has a **family history** of a genetic disorder they can have their **DNA analysed** to see if they have the **mutation** that causes it or if they are a **carrier** (see p. 44). If they're **tested** and **diagnosed before symptoms develop**, any **treatment** available can **begin earlier**. Also, knowing if they have the disorder or if they're a carrier can help to figure out if any **children** they have (or might have) are at **risk**.

Treatment
1) **Gene therapy** (see page 104) — this may be able to treat **some genetic disorders**. E.g. scientists have shown it's possible to **treat** symptoms of **cystic fibrosis** by inserting a **normal copy** of the **mutated gene**.
2) The **treatment** can be **different** for different **mutations** — for example, the **exact gene mutation** for Huntington's disease affects **symptom treatment options** as it affects the **time of onset** of symptoms.
3) **Early diagnosis** can affect treatment options — for example, if **sickle cell anaemia** is diagnosed at **birth**, treatments that **relieve symptoms** and work to **avoid complications** can be given straight away.

Prevention
Carriers or **sufferers** of genetic disorders can undergo **preimplantation genetic diagnosis** during *in vitro* **fertilisation** (IVF) to **prevent** any **offspring** having the disease. **Embryos** are produced by IVF and **screened** for the mutation. Only embryos **without the mutation** are **implanted** in the womb.

Practice Questions

Q1 Describe how knowing which specific mutation a cancer is caused by affects diagnosis.

Q2 Describe how knowing which specific mutation a genetic disorder is caused by affects treatment.

Exam Question

Q1 Discuss how knowing if a disorder is caused by a hereditary or acquired mutation affects prevention and diagnosis of the disorder.

[25 marks]

My genes need therapy — they've all got holes in the knees...

So whether a disorder is caused by an acquired or hereditary gene mutation has a pretty big effect on things like prevention, diagnosis and treatment. Make sure you understand the differences so you're not caught out in the exam.

UNIT 5: SECTION 3 — GENETICS

Stem Cells

Stem cells — they're the daddy of all cells, the big cheese, the top dog, and the head honcho. And here's why...

Stem Cells are Able to Mature into Any Type of Body Cell

1) **Multicellular organisms** are made up from many **different cell types** that are **specialised** for their function, e.g. liver cells, muscle cells, white blood cells.

2) **All** these specialised cell types originally came from **stem cells**.

3) Stem cells are **unspecialised** cells that can develop into **other types** of cell.

4) Stem cells divide to become **new** cells, which then become **specialised**.

5) All multicellular organisms have some form of stem cell.

6) Stem cells are found in the **embryo** (where they become all the **specialised cells** needed to form a **fetus**) and in **some adult tissues** (where they become **specialised** cells that need to be **replaced**, e.g. stem cells in the **bone marrow** can become **red blood cells**).

7) Stem cells that can mature (develop) into **any type** of **body cell** in an organism are called **totipotent cells**.

8) Totipotent stem cells in humans are **only present** in the **early life** of an **embryo**.

9) After this point the embryonic stem cells **lose** their ability to **specialise** into **all** types of cells, but can still become a **wide range** of cells.

10) Only a few **stem cells** remain in mature animals and they can only differentiate into a **few types** of cells.

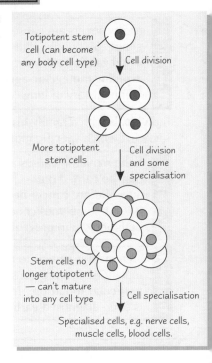

Totipotent stem cell (can become any body cell type)

Cell division

More totipotent stem cells

Cell division and some specialisation

Stem cells no longer totipotent — can't mature into any cell type

Cell specialisation

Specialised cells, e.g. nerve cells, muscle cells, blood cells.

Plants Contain Totipotent Stem Cells

1) Mature **plants** also have **stem cells** — they're found in areas where the plant is **growing**, e.g. in roots and shoots.

2) All stem cells in plants are **totipotent** — they can mature into any **cell type**.

3) This means they can be used to grow **plant organs** (e.g. roots) or **whole new plants** *in vitro* (artificially). Growing plant tissue artificially is called **tissue culture** (see next page).

Stem Cells Become Specialised Because Different Genes are Expressed

Totipotent stem cells become **specialised** because during their development they only **transcribe** and **translate part** of their **DNA**:

1) **Totipotent stem cells** all contain the **same genes** — but during **development** **not all** of them are **transcribed** and **translated** (expressed).

2) Under the **right conditions**, some **genes** are **expressed** and others are switched off.

3) **mRNA** is only **transcribed** from **specific genes**.

4) The mRNA from these genes is then **translated** into **proteins**.

5) These proteins **modify** the cell — they determine the cell **structure** and **control cell processes** (including the expression of **more genes**, which produces more proteins).

6) **Changes** to the cell produced by these proteins cause the cell to become **specialised**. These changes are **difficult** to **reverse**, so once a cell has specialised it **stays** specialised.

See pages 78-79 for more on transcription and translation.

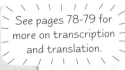

All of the girls expressed different jeans.

EXAMPLE: RED BLOOD CELLS

1) **Red blood cells** are produced from a type of **stem cell** in the **bone marrow**. They contain lots of **haemoglobin** and have **no nucleus** (to make room for more haemoglobin).

2) The stem cell produces a new cell in which the genes for **haemoglobin production** are **expressed**. Other genes, such as those involved in **removing the nucleus**, are **expressed** too. Many other genes are expressed or switched off, resulting in a specialised red blood cell.

Stem Cells

Tissue Culture Can be Used to Grow Plants from a Single Totipotent Cell

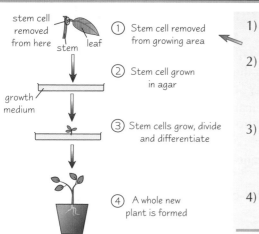

stem cell removed from here — stem — leaf — growth medium

1) Stem cell removed from growing area
2) Stem cell grown in agar
3) Stem cells grow, divide and differentiate
4) A whole new plant is formed

1) A **single totipotent stem cell** is taken from a **growing area** on a plant (e.g. a **root** or **shoot**).

See p. 66 for more on growth factors.

2) The cell is placed in some **growth medium** (e.g. agar) that contains **nutrients** and **growth factors**. The growth medium is **sterile**, so microorganisms can't grow and compete with the plant cells.

3) The plant stem cell will **grow** and **divide** into a **mass** of **unspecialised** cells. If the **conditions** are **suitable** (e.g. the plant cells are given the **right growth factors**), the cells will **mature** (develop) into **specialised** cells.

4) The cells grow and specialise to form a **plant organ** or an **entire plant** depending on the growth factors used.

You Need to be Able to Interpret Data About Tissue Culture

Here's an example of the kind of data that might crop up in the exam:

The **table** on the right shows the results of a **tissue culture experiment** — samples of **plant tissue** were taken from a **shoot** and grown on media with **varying ratios** of the growth factors **auxin** and **cytokinin**.

Growth medium	Ratio of auxin : cytokinin	Growth after 2 months
1	1 : 1	Growth but no specialised cells
2	1 : 25	Shoot formation
3	25 : 1	Root formation

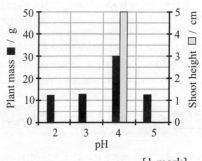

Jacob had probably overdone the growth factors for this one.

1) Growth medium **1** showed **no specialised cell growth** — so an **equal ratio** (1 : 1) of auxin : cytokinin promotes **unspecialised cell growth**.

2) Growth medium **2** showed **shoot formation** — so a **high cytokinin : auxin ratio** (25 : 1) promotes the growth of **specialised shoot cells**.

3) Growth medium **3** showed **root formation** — so a **high auxin : cytokinin ratio** (25 : 1) promotes the growth of **specialised root cells**.

The results of this experiment show that the **ratio** of these **growth factors** helps to control the **specialisation** of different tissues in this plant.

Practice Questions

Q1 What is a stem cell?

Q2 Where are stem cells found in plants?

Exam Question

Q1 Samples of plant tissue containing totipotent cells were taken from a stem and placed on growth media of different pH values. Each tissue sample weighed less than 1 g at the start of the experiment. The graph shows the results.
a) What are totipotent cells? [1 mark]
b) Describe and explain the results. [3 marks]

A tissue culture — what you need when you have a cold...

Jokes aside, all your biology knowledge is going to stem from some good old revision... Get it...? Stem... Sorry, I couldn't resist that one. But I mean it now — you need to know all about stem cells and how they become specialised to carry out a particular function. When you've got that straight, take a look at the tissue culture stuff. Plants are très exciting.

Stem Cells in Medicine

Like I said before, stem cells really are the daddy of all cells because they can divide and turn into all cell types. And it's this ability to turn into any other cell type that's got scientists and doctors fairly excited...

Some Stem Cell Therapies Already Exist

1) Stem cells can divide into **any cell type**, so they could be used to **replace** cells **damaged** by illness or injury.

2) Some stem cell therapies **already exist** for some diseases affecting the **blood** and **immune system**.

3) **Bone marrow** contains **stem cells** that can become specialised to form **any type** of **blood cell**. **Bone marrow transplants** can be used to replace the **faulty** bone marrow in patients that produce **abnormal blood cells**. The stem cells in the transplanted bone marrow **divide** and **specialise** to produce healthy blood cells.

4) This technique has been used successfully to treat **leukaemia** (a **cancer** of the blood or bone marrow) and **lymphoma** (a cancer of the **lymphatic system**).

5) It has also been used to treat some **genetic disorders**, such as **sickle-cell anaemia** and **severe combined immunodeficiency** (**SCID**):

> **Example**
>
> **Severe combined immunodeficiency** (**SCID**) is a genetic disorder that affects the immune system. People with SCID have a **poorly functioning immune system** as their **white blood cells** (made in the bone marrow from stem cells) are **defective**. This means they **can't defend** the body against infections by identifying and destroying microorganisms. So SCID sufferers are **extremely susceptible** to **infections**. Treatment with a **bone marrow transplant** replaces the faulty bone marrow with donor bone marrow that contains **stem cells without** the **faulty genes** that cause SCID. These then **differentiate** to produce **functional** white blood cells. These cells can identify and destroy invading pathogens, so the **immune system functions properly**.

Stem Cells Could be Used to Treat Other Diseases

Totipotent stem cells can develop into **any** specialised cell type, so scientists think they could be used to **replace damaged tissues** in a **range** of **diseases**. Scientists are **researching** the use of stem cells as **treatment** for lots of conditions, including:

- **Spinal cord injuries** — stem cells could be used to replace damaged **nerve tissue**.
- **Heart disease** and **damage caused by heart attacks** — stem cells could be used to replace damaged **heart tissue**.
- **Bladder conditions** — stem cells could be used to grow **whole bladders**, which are then **implanted** in patients to replace diseased ones.
- **Respiratory diseases** — **donated windpipes** can be stripped down to their simple collagen structure and then covered with **tissue** generated by stem cells. This can then be **transplanted** into patients.
- **Organ transplants** — organs could be **grown** from stem cells to provide new organs for people on **donor waiting lists**.

These treatments aren't available yet but some are in the early stages of clinical trials.

It might even be possible to make **stem cells genetically identical** to a **patient's own cells**. These could then be used to **grow** some **new tissue** or **an organ** that the patient's body **wouldn't reject** (rejection of transplants occurs quite often and is caused by the patient's immune system recognising the tissue as **foreign** and **attacking it**).

There are Huge Benefits to Using Stem Cells in Medicine

People who make **decisions** about the **use** of stem cells to treat human disorders have to consider the **potential benefits** of stem cell therapies:

- They could **save** many **lives** — e.g. many people waiting for organ transplants **die** before a **donor organ** becomes available. Stem cells could be used to **grow organs** for those people awaiting transplants.
- They could **improve** the **quality of life** for many people — e.g. stem cells could be used to replace damaged cells in the eyes of people who are **blind**.

Stem Cells in Medicine

Human *Stem Cells* Can Come from *Adult Tissue* or *Embryos*

To **use stem cells** scientists have to get them from somewhere. There are **two** potential **sources** of human stem cells:

(1) Adult stem cells

1) These are obtained from the **body tissues** of an **adult**. For example, adult stem cells are found in **bone marrow**.

2) They can be obtained in a relatively **simple operation** — with very **little risk** involved, but quite **a lot** of **discomfort**.

3) Adult stem cells **aren't** as **flexible** as embryonic stem cells — they can only specialise into a **limited** range of cells, not all body cell types. Although scientists are **trying** to find ways to make adult stem cells **specialise** into **any cell type**.

(2) Embryonic stem cells

1) These are obtained from **embryos** at an **early stage of development**.

2) Embryos are created in a **laboratory** using *in vitro* fertilisation (IVF) — **egg cells** are **fertilised** by sperm **outside the womb**.

3) Once the embryos are approximately **4 to 5 days old**, **stem cells** are **removed** from them and the rest of the embryo is **destroyed**.

4) Embryonic stem cells can develop into **all types** of specialised cells.

There are *Ethical Issues* Surrounding Stem Cell Use

1) Obtaining stem cells from **embryos** created by IVF raises **ethical issues** because the procedure results in the **destruction** of an embryo that **could** become a fetus if placed in a **womb**.

2) Some people believe that at the moment of **fertilisation** an **individual** is formed that has the **right** to **life** — so they believe that it's **wrong** to **destroy** embryos.

3) Some people have **fewer objections** to stem cells being **obtained** from **unfertilised embryos** — embryos made from **egg cells** that **haven't** been fertilised by sperm. This is because the embryos **couldn't survive** past a few days and **wouldn't** produce a fetus if placed in a womb.

4) Some people think that **scientists** should **only use** adult stem cells because their production **doesn't** destroy an embryo. But adult stem cells **can't** develop into all the specialised cell types that embryonic stem cells can.

5) The decision makers in **society** have to take into account **everyone's views** when making decisions about **important scientific work** like stem cell research and its use to treat human disorders.

Practice Questions

Q1 What types of cells can bone marrow stem cells produce?

Q2 Name two conditions that stem cells could potentially be used to treat.

Q3 Describe one difference between embryonic and adult stem cells.

Exam Questions

Q1 Explain one way in which stem cell therapy is currently being used. [4 marks]

Q2 Explain why some people object to the use of embryonic stem cells in treating human disorders. [2 marks]

It's OK — you can grow yourself a new brain especially for this revision...

And that's the end of this section — whoopdidoo. It was a whopper — in size, I mean, not as in the famous 1980s chewy bars. Before you zoom off to something else (because I know you can't wait), take the time to learn all of the pages, including these last two. It might take a little while, but you'll be glad of it when it comes to the exam. Promise.

Making DNA Fragments

You might have just done a section about genetics, but the good stuff's all in here. Three, two, one... go.

Gene Technology — Techniques Used to Study Genes

Gene technology is basically all the **techniques** that can be used to **study genes** and their **function**. Examples include:

- The **polymerase chain reaction** (**PCR**) — produces **lots** of **identical copies** of a specific **gene** (see next page).
- *In vivo* gene cloning — also produces **lots** of **identical copies** of a specific **gene** (see page 94).
- **DNA probes** — used to **identify** specific genes (see page 100).

Scientists use these techniques to do many things (as well as study genes) — e.g. they use them for **genetic engineering** (see p. 96), **DNA fingerprinting** (see p. 98), **diagnosing diseases** (see p. 99) and **treating genetic disorders** (see p. 103).

Gene Technology Uses DNA Fragments

As gene technology is all about **studying genes**, a good place to start is learning how to **get a copy** of the **DNA fragment** containing the gene you're **interested** in (the **target gene**). There are **three ways** that DNA fragments can be produced:

① Using Reverse Transcriptase

1) Many **cells** only contain **two copies** of each gene, making it **difficult** to obtain a DNA fragment containing the target gene. But they can contain **many mRNA** molecules (see p. 78) complementary to the gene, so mRNA is often **easier** to obtain.

2) The mRNA molecules can be used as **templates** to **make lots of DNA**. The **enzyme reverse transcriptase makes DNA** from an RNA template. The DNA produced is called **complementary DNA (cDNA)**.

3) For example, **pancreatic cells** produce the protein **insulin**. They have loads of mRNA molecules complementary to the **insulin gene**, but only **two copies** of the gene **itself**. So reverse transcriptase could be used to **make cDNA** from the **insulin mRNA**.

4) To do this, **mRNA** is first isolated from cells. Then it's **mixed** with **free DNA nucleotides** and **reverse transcriptase**. The reverse transcriptase uses the mRNA as a **template** to synthesise a **new strand** of cDNA.

② Using Restriction Endonuclease Enzymes

1) Some sections of DNA have **palindromic** sequences of **nucleotides**. These sequences consist of **antiparallel base pairs** (base pairs that read the **same** in **opposite directions**).

2) **Restriction endonucleases** are enzymes that **recognise specific** palindromic sequences (known as **recognition sequences**) and **cut** (**digest**) the DNA at these places.

3) Different restriction endonucleases cut at **different specific** recognition sequences, because the **shape** of the recognition sequence is **complementary** to an enzyme's **active site**. E.g. the restriction endonuclease *Eco*RI cuts at GAATTC, but *Hind*III cuts at AAGCTT.

4) If recognition sequences are present at **either side** of the DNA fragment you want, you can use restriction endonucleases to **separate** it from the rest of the DNA.

5) The DNA sample is **incubated** with the specific restriction endonuclease, which **cuts** the DNA fragment out via a **hydrolysis reaction**.

6) Sometimes the cut leaves **sticky ends** — **small tails** of **unpaired bases** at **each end** of the fragment. Sticky ends can be used to **bind** (**anneal**) the DNA fragment to another piece of DNA that has sticky ends with **complementary sequences** (there's more about this on p. 94).

Making DNA Fragments

3) Using the **Polymerase Chain Reaction (PCR)**

The **polymerase chain reaction** (PCR) can be used to make **millions of copies** of a fragment of DNA in just a few hours. PCR has **several stages** and is **repeated** over and over to make lots of copies:

1) A reaction mixture is set up that contains the **DNA sample, free nucleotides, primers** and **DNA polymerase**.
   - **Primers** are short pieces of DNA that are **complementary** to the bases at the **start** of the fragment you want.
   - **DNA polymerase** is an **enzyme** that creates new DNA strands.

2) The DNA mixture is **heated** to **95 °C** to break the **hydrogen bonds** between the two strands of DNA.

3) The mixture is then **cooled** to between **50** and **65 °C** so that the primers can **bind** (anneal) to the strands.

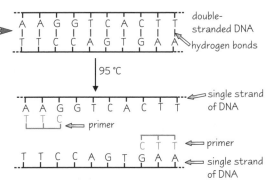

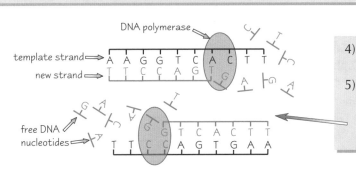

4) The reaction mixture is heated to **72 °C**, so **DNA polymerase** can **work**.

5) The DNA polymerase **lines up** free DNA nucleotides **alongside** each **template strand**. Specific **base pairing** means **new complementary strands** are formed.

> PCR produces loads of identical copies of DNA, so it can be used to clone genes — this is called *in vitro* cloning.

6) **Two new copies** of the fragment of DNA are formed and **one cycle** of PCR is **complete**.

7) The cycle starts again, with the mixture being heated to 95 °C and this time **all four strands** (two original and two new) are used as **templates**.

8) Each PCR cycle **doubles** the amount of DNA, e.g. **1st cycle** = 2 × 2 = **4 DNA fragments**, **2nd cycle** = 4 × 2 = **8 DNA fragments**, **3rd cycle** = 8 × 2 = **16 DNA fragments**, and so on.

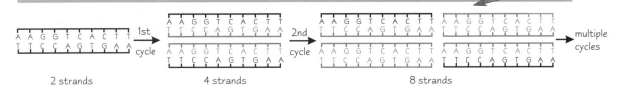

Practice Questions

Q1 Name three ways a DNA fragment can be produced.
Q2 What is reverse transcriptase?
Q3 What are sticky ends?
Q4 What does PCR stand for?
Q5 What is a primer?

```
G G A T C C G T T T C A G G A T C C
C C T A G G C A A A G T C C T A G G
```
DNA fragment wanted

Exam Question

Q1 A fragment of DNA (shown above) needs to be isolated from some bacterial DNA. The restriction endonuclease *Bam*HI recognises the sequence GGATCC and cuts between G and G.
   a) Explain how *Bam*HI could be used to isolate the DNA fragment. [2 marks]
   b) Describe and explain how to produce multiple copies of this DNA fragment using PCR. [5 marks]

Sticky ends — for once a name that actually makes sense.

Okay, your eyes might have gone funny from seeing so many nucleotides on these two pages. But once you've recovered, it's really important to go over these pages as many times as you need to, 'cause examiners love throwing in a few questions on restriction enzymes or PCR. Bless 'em — examiners get excited about the strangest things.

Gene Cloning

Hmmmmm... nope, can't think of any exciting or funny ways to start this double page. Sorry.

Gene Cloning can be done In Vitro or In Vivo

In vitro is Latin for within glass. In vivo is Latin for within the living.

Gene cloning is all about making loads of **identical copies** of a gene.
This can be done using **two** different techniques:

1) *In vitro* cloning — where the gene copies are made **outside** of a **living organism** using **PCR** (see page 93).

2) *In vivo* cloning — where the gene copies are made **within** a **living organism**. As the organism **grows** and **divides**, it **replicates** its **DNA**, creating multiple copies of the gene (see below).

In Vivo Cloning Step 1 — The Gene is Inserted into a Vector

The **DNA fragment** containing the **target gene** has been isolated using one of the techniques on pages 92-93. The first step in *in vivo* cloning is to **stick** the fragment **into** a **vector** using **restriction endonuclease** and **ligase** (an enzyme):

1) The DNA fragment is inserted into vector DNA — a **vector** is something that's used to **transfer DNA** into a **cell**. They can be **plasmids** (**small, circular molecules** of DNA in **bacteria**) or **bacteriophages** (**viruses** that **infect** bacteria).

2) The vector DNA is **cut open** using the **same** restriction endonuclease that was used to **isolate** the DNA fragment containing the target gene (see p. 92). So the **sticky ends** of the vector are **complementary** to the sticky ends of the DNA fragment containing the gene.

3) The vector DNA and DNA fragment are **mixed together** with **DNA ligase**. DNA ligase **joins** the sticky ends of the DNA fragment to the sticky ends of the vector DNA. This process is called **ligation**.

4) The new combination of bases in the DNA (vector DNA + DNA fragment) is called **recombinant DNA**.

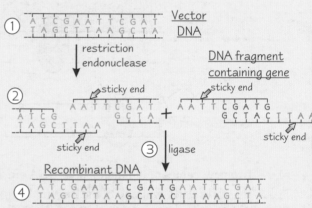

In Vivo Cloning Step 2 — The Vector Transfers the Gene into Host Cells

1) The **vector** with the **recombinant DNA** is used to **transfer** the gene into **cells** (called **host** cells).

2) If a **plasmid vector** is used, **host cells** have to be **persuaded** to **take in** the plasmid vector and its DNA. E.g. host bacterial cells are placed into ice-cold **calcium chloride** solution to make their cell walls more **permeable**. The **plasmids** are **added** and the mixture is **heat-shocked** (heated to around **42 °C** for **1-2 minutes**), which encourages the cells to take in the plasmids.

3) With a **bacteriophage** vector, the bacteriophage will **infect** the host bacterium by **injecting** its **DNA** into it. The phage DNA (with the target gene in it) then **integrates** into the bacterial DNA.

4) **Host cells** that **take up** the vectors containing the gene of interest are said to be **transformed**.

In Vivo Cloning Step 3 — Identifying Transformed Host Cells

Not all host cells will have **taken up** the vector and its DNA. **Marker genes** can be used to **identify** the **transformed cells**

1) **Marker genes** can be inserted into vectors at the **same time** as the gene to be cloned. This means any **transformed host cells** will contain the gene to be cloned **and** the marker gene.

2) Host cells are **grown** on **agar plates** and each cell **divides** and **replicates** its DNA, creating a **colony** of **cloned cell**

3) Transformed cells will produce colonies where **all the cells** contain the cloned gene and the marker gene.

4) The marker gene can code for **antibiotic resistance** — host cells are grown on agar plates **containing** the specific **antibiotic**, so **only** transformed cells that have the **marker gene** will **survive** and **grow**.

5) The marker gene can code for **fluorescence** — when the agar plate is placed under a **UV light only** transformed cells will **fluoresce**.

6) **Identified** transformed cells are allowed to **grow more**, producing **lots** and **lots** of **copies** of the **cloned gene**.

Gene Cloning

There are **Advantages** and **Disadvantages** to Both **In Vivo** and **In Vitro** Cloning

Depending on the **reason** why you want to clone a gene, you can choose to do it either by *in vivo* cloning or *in vitro* cloning. Both techniques have **advantages** and **disadvantages**:

In Vivo Cloning

Either the cloning had worked, or Professor Dim's eyesight had gone.

1) Cloning *in vivo* can produce **mRNA** and **protein** as well as DNA because it's done in a **living cell** (which has the ribosomes and all the enzymes needed to produce them).

2) Cloning *in vivo* can also produce **modified DNA**, **modified mRNA** or **modified protein** — they have **modifications added** to them, e.g. **sugar** or **methyl** (-CH₃) groups.

3) **Large fragments** of DNA can be cloned using *in vivo* cloning, e.g. between **20 to 45 kilobases** of DNA can be inserted into some **plasmids** and **bacteriophages**.

4) *In vivo* cloning can be a **relatively cheap method**, depending on **how much** DNA you want to produce.

In vivo cloning also has **disadvantages** — the DNA fragment has to be **isolated** from other cell components, you may **not want modified** DNA, and it can be quite a **slow** process (because some types of bacteria **grow quite slowly**).

> A kilobase is 1000 nucleotide bases. It's often shortened to kb.

In Vitro Cloning (PCR)

1) *In vitro* cloning can be used to produce **lots** of DNA (but **not** mRNA or protein).

2) The DNA produced **isn't modified** (see above) — an advantage if you **don't want** it to be modified.

3) This technique **only** replicates the **DNA fragment** of **interest** (e.g. the target gene). This means that you **don't** have to **isolate** the DNA fragment from **host DNA** or **cell components**.

4) *In vitro* cloning is a **fast** process — PCR can clone **millions of copies** of DNA in just a **few hours**.

In vitro cloning also has **disadvantages** — it can only replicate a **small DNA fragment** (compared to *in vitro*), you may want a **modified** product, **mRNA** and **protein** aren't made as well, and it can be **expensive** if you want to produce a lot of DNA.

Practice Questions

Q1 What is *in vitro* gene cloning?

Q2 What is a vector?

Q3 Other than a plasmid, give an example of a vector.

Q4 Name the type of enzyme that can be used to cut DNA.

Q5 What is the name of the type of DNA formed from vector DNA and an inserted DNA fragment?

Q6 What is a marker gene?

Q7 Give two advantages of *in vivo* cloning.

Q8 *In vitro* cloning is a slow process — is this statement correct?

Agar plate under UV light

agar plate

colony A

colony B

colony not visible ●

colony visible ○

Exam Question

Q1 A scientist has cloned a gene by transferring a plasmid containing the target gene and a fluorescent marker gene into some bacterial cells. The cells were grown on an agar plate. The plate was then placed under UV light (see above).
   a) Explain why the scientist thinks colony A contains transformed host cells, but colony B doesn't. [2 marks]
   b) Explain how the scientist might have inserted the target gene into the plasmid. [3 marks]

Transformed boyfriends — made to listen, tidy up and agree with you...

This page is quite scary I know. But don't worry, it's not as difficult as photosynthesis or respiration — you just need to keep going over the steps of all these different techniques until they make sense. And they do make sense really, promise. I know I've said it before, but drawing out the diagram will help — then you'll know inserting DNA into vectors like a pro.

Genetic Engineering

Now that you know how to make a DNA fragment and clone a gene, it's probably a good time to tell you why you might want to. Don't worry — it's not evil stuff, but I promise to do my evil laugh. Mwah ha hah.

Genetic Engineering is the Manipulation of an Organism's DNA

1) **Genetic engineering** is also known as **recombinant DNA technology**.

2) Organisms that have had their **DNA altered** by genetic engineering are called **transformed organisms**.

Transformed organisms are also known as genetically engineered or genetically modified organisms.

3) These organisms have **recombinant DNA** — **DNA** formed by **joining together** DNA from **different sources**.

4) **Microorganisms**, **plants** and **animals** can all be **genetically engineered** to **benefit humans**.

5) **Transformed microorganisms** can be made using the same technology as *in vivo* cloning (see page 94). For example, **foreign DNA** can be **inserted** into **microorganisms** to produce **lots** of **useful protein**, e.g. insulin:

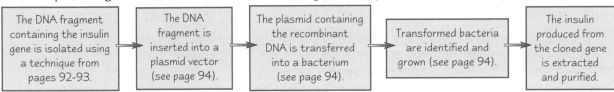

| The DNA fragment containing the insulin gene is isolated using a technique from pages 92-93. | → | The DNA fragment is inserted into a plasmid vector (see page 94). | → | The plasmid containing the recombinant DNA is transferred into a bacterium (see page 94). | → | Transformed bacteria are identified and grown (see page 94). | → | The insulin produced from the cloned gene is extracted and purified. |

6) **Transformed plants** can also be produced — a gene that codes for a **desirable characteristic** is inserted into a **plasmid**. The plasmid is added to a **bacterium** and the bacterium is used as a **vector** to get the gene into the **plant cells**. The transformed plant will have the desirable characteristic coded for by that gene.

7) **Transformed animals** can be produced too — a gene that codes for a **desirable characteristic** is inserted into an **animal embryo**. The transformed animal will have the desirable characteristic coded for by that gene.

Transformed (Genetically Engineered) Organisms can Benefit Humans

Producing **transformed organisms** (microorganisms, plants and animals) can benefit **humans** in lots of ways:

1 Agriculture

- **Agricultural crops** can be **transformed** so that they give **higher yields** or are **more nutritious**. This means these plants can be used to reduce the risk of **famine** and **malnutrition**. Crops can also be transformed to have **pest resistance**, so that **fewer pesticides** are needed. This **reduces costs** and reduces any **environmental problems** associated with using pesticides.

- For example, *Golden Rice* is a variety of **transformed rice**. It contains **one gene** from a **daffodil plant** and **one gene** from a **soil bacterium**, which together enable the rice to produce **beta-carotene**. The beta-carotene is used by our bodies to produce **vitamin A**. *Golden Rice* is being developed to **reduce vitamin A deficiency** in areas where there's a **shortage** of **dietary vitamin A**, e.g. **south Asia**, **Africa**. Vitamin A deficiency is a big problem in these areas, e.g. up to **500 000 children per year** worldwide go **blind** due to vitamin A deficiency.

2 Industry

- **Industrial processes** often use **biological catalysts** (**enzymes**). These enzymes can be produced from **transformed organisms**, so they can be produced in **large quantities** for **less money**, **reducing costs**.

- For example, **chymosin** (or **rennin**) is an enzyme used in **cheese-making**. It used to be made from **rennet** (a substance produced in the **stomach** of **cows**), but it can now be produced by **transformed organisms**. This means it can be made in **large quantities**, relatively **cheaply** and **without killing** any **cows**, making some cheese suitable for **vegetarians**.

3 Medicine

- Many **drugs** and **vaccines** are produced by transformed organisms, using recombinant DNA technology. They can be made **quickly**, **cheaply** and in **large quantities** using this method.

- For example, **insulin** is used to treat **Type 1 diabetes** and used to come from **animals** (cow, horse or pig pancreases). This insulin **wasn't** human insulin though, so it **didn't work quite as well**. Human insulin is now made from **transformed microorganisms**, using a **cloned human insulin gene** (see above).

Genetic Engineering

Many People are **Concerned** About the Use of **Genetic Engineering**

There are **ethical**, **moral** and **social concerns** associated with the **use** of **genetic engineering**:

1 Agriculture

- **Farmers** might plant only **one type** of transformed crop (this is called **monoculture**). This could make the **whole crop vulnerable** to **disease** because the plants are **genetically identical**.
- Some people are concerned about the possibility of **'superweeds'** — weeds that are **resistant** to **herbicides**. These could occur if transformed crops **interbreed** with **wild plants**.

2 Industry

- **Without proper labelling**, some people think they **won't** have a **choice** about whether to consume food made using genetically engineered organisms.
- Some people are worried that the process used to **purify** proteins (from genetically engineered organisms) could lead to the introduction of **toxins** into the **food industry**.

3 Medicine

- Companies who **own** genetic engineering technologies may **limit** the **use** of technologies that could be **saving lives**.
- Some people worry this technology could be used **unethically**, e.g. to make **designer babies** (babies that have characteristics **chosen** by their parents). This is currently **illegal** though.

Humanitarians Think Genetic Engineering will Benefit People

Genetic engineering has **many** potential **humanitarian benefits**:

1) **Agricultural crops** could be produced that help **reduce** the risk of **famine** and **malnutrition**, e.g. **drought-resistant** crops for **drought-prone** areas.
2) **Transformed crops** could be used to produce **useful pharmaceutical products** (e.g. **vaccines**) which could make drugs **available** to **more people**, e.g. in areas where **refrigeration** (usually needed for **storing** vaccines) **isn't available**.
3) **Medicines** could be produced more **cheaply**, so more people can **afford** them.

You need to be able to balance the humanitarian benefits with opposing views from environmentalists and anti-globalisation activists.

But some **environmentalists** and **anti-globalisation activists** have concerns:

1) **Environmentalists** — Many **oppose** recombinant DNA technology because they think it could **potentially damage** the **environment**. E.g. transformed crops could encourage **farmers** to carry out monoculture (see above), which **decreases biodiversity**. There are also fears that if **transformed crops breed** with **wild plants** there'll be **uncontrolled spread** of **recombinant DNA**, with **unknown consequences**.
2) **Anti-globalisation activists** — These are people who **oppose** globalisation (e.g. the **growth of large multinational companies** at the **expense** of **smaller ones**). A few, **large** biotechnology companies **control** some forms of genetic engineering. As the **use** of this technology **increases**, these companies get **bigger** and **more powerful**. This may **force** smaller companies **out of business**, e.g. by making it **harder** for them to **compete**.

Practice Questions

Q1 What is recombinant DNA?
Q2 Give an example of a transformed agricultural crop.

Exam Question

Q1 A large agricultural company has isolated a gene from bacteria that may increase the drought resistance of wheat plants.
   a) Briefly explain how this gene could be used to make a transformed wheat plant. [3 marks]
   b) Suggest how the transformed wheat plants might be beneficial to humans. [2 marks]
   c) Suggest why anti-globalisation activists may be against the use of this gene. [1 mark]

Neapolitan — recombinant ice cream...

Ahhh, sitting in the sun, licking an ice cream, exams all over. That's where you'll be in a few months' time. After revising all this horrible stuff that is. As genetic engineering advances, more questions will pop up about its implications. So it's a good idea to know all sides of the argument — you need to know them for the exam anyway.

Genetic Fingerprinting

100 years ago they were starting to identify people using their fingerprints, but now we can use their DNA instead.

Genomes *Contain* Repetitive, Non-Coding DNA Sequences

1) **Not all** of an organism's **genome** (all the genetic material in an organism) **codes** for **proteins**.

2) Some of the genome consists of **repetitive**, **non-coding base sequences** — base sequences that **don't** code for proteins and **repeat** next to each other over and over (sometimes thousands of times), e.g. CATGCATGCATGCATG is a repeat of the non-coding base sequence CATG.

3) The **number of times** these sequences are **repeated differs** from person to person, so the **length** of these sequences in nucleotides differs too. E.g. a **four** nucleotide sequence might be repeated **12 times** in one person = **48 nucleotides** (12 × 4), but repeated **16 times** in another person = **64 nucleotides** (16 × 4).

4) The repeated sequences occur in **lots of places** in the **genome**. The **number** of times a **sequence is repeated** (and so the number of nucleotides) at **different places** in their genome can be **compared** between **individuals** — this is called **genetic fingerprinting**.

5) The **probability** of **two individuals** having the **same** genetic fingerprint is **very low** because the **chance** of **two individuals** having the **same number** of sequence repeats at **each place** they're found in DNA is **very low**.

Electrophoresis *Separates* DNA Fragments *to Make a* Genetic Fingerprint

So **genetic fingerprints** can be **compared** between **different individuals**. Now you need to know how one is **made**:

1) A **sample** of **DNA** is obtained, e.g. from a person's **blood**, **saliva** etc.

2) **PCR** (see page 93) is used to make **many copies** of the **areas** of DNA that contain the repeated sequences — **primers** are used that bind to **either side** of these **repeats** and so the **whole** repeat is amplified.

3) You end up with **DNA fragments** where the **length** (in nucleotides) corresponds to the **number of repeats** the person has at each specific position, e.g. one person may have 80 nucleotides, another person 120.

4) A **fluorescent tag** is added to all the DNA fragments so they can be viewed under **UV light**.

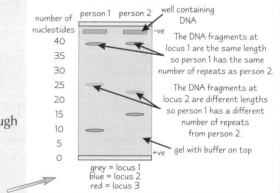

Genetic Fingerprinting

5) The DNA fragments undergo **electrophoresis**:

- The DNA mixture is placed into a **well** in a slab of **gel** and covered in a **buffer solution** that **conducts electricity**.

- An **electrical current** is passed through the gel — DNA fragments are **negatively charged**, so they **move towards** the **positive electrode** at the far end of the gel.

- **Small** DNA fragments move **faster** and **travel further** through the gel, so the DNA fragments **separate** according to **size**.

6) The DNA fragments are viewed as **bands** under **UV light** — this is the **genetic fingerprint**.

7) Two genetic fingerprints can be **compared** — e.g. if both fingerprints have a band at the **same location** on the **gel** it means they have the **same number** of **nucleotides** and so the **same number** of **sequence repeats** at that place — it's a **match**.

Genetic Fingerprinting *is Used to* Determine Relationships *and* Variability

Genetic fingerprinting has **many uses**, which include:

- **Determining genetic relationships** — We **inherit** the repetitive, non-coding base sequences from our **parents**. Roughly **half** of the sequences come from **each parent**. This means the **more bands** on a genetic fingerprint that match, the more **closely related** (**genetically similar**) two people are. E.g. **paternity tests** are used to determine the **biological father** of a child by comparing genetic fingerprints. If lots of bands on the fingerprint **match**, then that person is **most probably** the child's father. The **higher** the **number** of places in the genome compared, the more **accurate** the test result.

- **Determining genetic variability within a population** — The **greater** the **number of bands** that **don't** match on a genetic fingerprint, the more **genetically different** people are. This means you can **compare** the **number of repeats** at **several places** in the genome for a population to find out how **genetically varied** that population is. E.g. the **more** the **number of repeats** varies at **several places**, the **greater** the **genetic variability** within a population.

Genetic Fingerprinting

Genetic Fingerprinting can be Used in Forensic Science...

Forensic scientists use genetic fingerprinting to **compare** samples of **DNA** collected from **crime scenes** (e.g. DNA from **blood**, **semen**, **skin cells**, **saliva**, **hair** etc.) to samples of DNA from **possible suspects**, to **link them** to crime scenes.

1) The **DNA** is **isolated** from all the collected samples (from the crime scene and from the suspects).
2) Each sample is **replicated** using **PCR** (see p. 93).
3) The **PCR products** are run on an **electrophoresis gel** and the genetic fingerprints produced are **compared** to see if any match.
4) If the samples match, it **links** a **person** to the **crime scene**. E.g. this gel shows that the genetic fingerprint from **suspect C** matches that from the crime scene, **linking** them to the crime scene. All five bands match, so suspect C has the **same number** of repeats (nucleotides) at **five** different places.

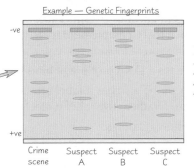

Example — Genetic Fingerprints

-ve / +ve

Crime scene | Suspect A | Suspect B | Suspect C

PCR amplifies the DNA, so enough is produced for it to be seen on the gel.

...Medical Diagnosis...

- In medical diagnosis, a genetic fingerprint can refer to a **unique pattern** of **several alleles**.
- It can be used to **diagnose genetic disorders** and **cancer**. It's useful when the **specific** mutation **isn't** known or where **several mutations** could have caused the disorder, because it identifies a **broader**, **altered** genetic pattern.

  EXAMPLE: **Preimplantation genetic haplotyping** (**PGH**) **screens embryos** created by **IVF** for genetic disorders **before** they're **implanted** into the uterus. The **faulty regions** of the **parents' DNA** are used to produce **genetic fingerprints**, which are **compared** to the genetic fingerprint of the **embryo**. If the fingerprints **match**, the embryo has **inherited** the **disorder**. It can be used to screen for **cystic fibrosis**, **Huntington's disease** etc.

  EXAMPLE: Genetic fingerprinting can be used to **diagnose sarcomas** (types of **tumour**). Conventional methods of identifying a tumour (e.g. biopsies) only show the **physical differences** between tumours. Now the **genetic fingerprint** of a known sarcoma (e.g. the **different mutated alleles**) can be **compared** to the genetic fingerprint of a **patient's tumour**. If there's a **match**, the sarcoma can be specifically **diagnosed** and the **treatment** can be targeted to that specific type (see page 103).

  A specific mutation can be found using gene probes and sequencing (see p. 100-101).

...and Animal and Plant Breeding

Genetic fingerprinting can be used on **animals** and **plants** to **prevent inbreeding**, which causes **health**, **productivity** and **reproductive problems**. Inbreeding **decreases** the **gene pool** (the number of **different alleles** in a population, see p. 46), which can lead to an **increased risk** of **genetic disorders**, leading to **health problems** etc. Genetic fingerprinting can be used to **identify** how **closely-related** individuals are — the **more closely-related** two individuals are, the **more similar** their genetic fingerprint will be (e.g. **more bands** will **match**). The **least related** individuals will be **bred together**.

Practice Questions

Q1 Why are two people unlikely to have the same genetic fingerprint?
Q2 In gel electrophoresis, which electrode do DNA fragments move towards?
Q3 Why might genetic fingerprinting be used in forensic science?

-ve / +ve

Child | 1 | 2

Exam Question

Q1 The diagram on the right shows three genetic fingerprints — one from a child and two from possible fathers.
  a) Describe how the genetic fingerprint is made. [5 marks]
  b) Which genetic fingerprint is most likely to be from the child's father? Explain your answer. [2 marks]
  c) Give another use of genetic fingerprint technology. [1 mark]

Fingerprinting — in primary school it involved lots of paint and paper...

Who would have thought that tiny pieces of DNA on a gel would be that important? Well, they are and you need to know all about them. Make sure you know the theory behind fingerprinting as well as its applications. And remember, it's very unlikely that two people will have the same genetic fingerprint (except identical twins that is).

Locating and Sequencing Genes

We're gonna take gene technology to a whole new level now, with restriction mapping and DNA sequencing.

You can **Look** for **Genes** Using **DNA Probes** and **Hybridisation**

1) DNA probes can be used to **locate genes** (e.g. on **chromosomes**) or see if a person's DNA **contains** a **mutated gene** (e.g. a gene that causes a **genetic disorder**).

2) **DNA** (gene) **probes** are **short strands** of **DNA**. They have a **specific base sequence** that's **complementary** to the base sequence of part of a **target gene** (the gene you're looking for, e.g. a gene that causes a genetic disorder).

3) This means a DNA probe will **bind** (**hybridise**) to the **target gene** if it's **present** in a **sample** of DNA.

4) A DNA probe also has a **label attached**, so that it can be **detected**. The two most common types of label are a **radioactive** label (detected using **X-ray film**) or a **fluorescent** label (detected using **UV light**).

5) Here's how it's done:

- A **sample** of **DNA** is **digested** into fragments using **restriction enzymes** (see page 92) and separated using electrophoresis (see page 98).
- The separated DNA fragments are then transferred to a **nylon membrane** and **incubated** with the **fluorescently labelled DNA probe**.
- If the gene is **present**, the DNA probe will **hybridise** (**bind**) to it.
- The **membrane** is then **exposed** to **UV light** and if the gene is present there will be a **fluorescent band**. E.g. **sample 3** has a visible band, so this patient has the **gene**.

The **Base Sequence** of a Gene can be **Determined** by **Restriction Mapping...**

As well as locating a **gene**, you might also want to know its **sequence** — this is done by **DNA sequencing** (see next page). But most genes are **too long** to be sequenced **all in one go**, so they're **cut** into **smaller sections** using **restriction enzymes**, then the smaller parts are sequenced. These smaller sections are then put back in the **correct order**, so the **entire gene sequence** can be **read** in the **right order** — restriction mapping can be used to do this:

1) **Different restriction enzymes** are used to **cut** labelled DNA into fragments (see page 92).

2) The DNA fragments are then **separated** by **electrophoresis** (see page 98).

3) The size of the **fragments** produced is used to **determine** the **relative locations** of **cut sites**.

4) A **restriction map** of the **original DNA** is made — a **diagram** of the piece of **DNA** showing the **different cut sites**, and so where the recognition sites of the restriction enzymes used are found.

Here's an example:

1) Some DNA, **10 kilobases** long (1 kb = **1000 nucleotides**) was **radioactively labelled**.

2) The DNA was **digested** using **two restriction enzymes**, **Hind**III and **Eco**RI, and the digested fragments were **separated** using **electrophoresis**.

3) The gel was used to build up a **restriction map** of the original DNA:

a) The gel shows that the DNA was cut into **two fragments** by **Hind**III, so there's **one Hind**III **recognition sequence** in one of two places. But because the **2 kb piece** is **radioactive**, the label must be on the 2 kb piece. So the **Hind**III site must be 2 kb from the label.

b) The gel shows that the DNA was cut into **two fragments** both **5 kb** long by **Eco**RI, so there's **one Eco**RI **recognition sequence** in the **middle** of the piece.

c) Finally, putting both of these together, the **complete restriction map** must be:

4) The restriction map matches the fragments of the **total digest** (where both enzymes are present and the DNA is cut at **all** of the **recognition sequences** present) — the **radioactive label** is on the 2 kb **Hind**III piece.

5) A **partial digest** is where the restriction enzymes **haven't** been **left long enough** to **cut** at all of their **recognition sequences**, producing **fragments of other lengths**, e.g. if *Eco*RI **doesn't** cut there'll be an **8 kb** fragment produced.

Locating and Sequencing Genes

... and Gene Sequencing

Gene sequencing is used to determine the **order** of **bases** in a section of **DNA** (gene). It can be carried out by the **chain termination method**, which lets you sequence **small fragments** of DNA, up to **750 base pairs**:

1) The following mixture is added to **four separate** tubes:
   - A **single-stranded DNA template** — the DNA to be sequenced.
   - **DNA polymerase** — the enzyme that joins DNA nucleotides together.
   - Lots of **DNA primer** — short pieces of DNA (see p. 93).
   - **Free nucleotides** — lots of free A, T, C and G nucleotides.
   - **Fluorescently-labelled modified nucleotide** — like a regular nucleotide, but once it's added to a DNA strand, **no more** bases can be added after it. A **different** modified nucleotide is added to **each tube** (A*, T*, C*, G*).

Jane and her friends had already made their costumes for the sequin-cing lesson.

2) The tubes undergo **PCR**, which produces many **strands of DNA**. The strands are **different lengths** because each one **terminates** at a **different point** depending on where the modified nucleotide was added.

3) For example, in tube A (with the **modified adenine** nucleotide A*) sometimes A* is **added** to the DNA at point 4 **instead** of A, **stopping** the **addition** of any more bases (the strand is **terminated**). Sometimes A is added at point 4, then A* is added at **point 5**. Sometimes A is added at **point 4**, A again at point 5, G at point 6 and A* is added at **point 7**. So strands of **three different lengths** (4 bases, 5 bases and 7 bases) all ending in A* are produced.

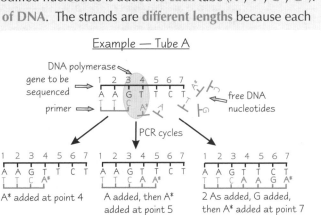

Example — Tube A

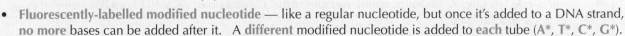

A* added at point 4   A added, then A* added at point 5   2 As added, G added, then A* added at point 7

4) The DNA fragments in each tube are separated by **electrophoresis** and **visualised** under **UV light** (because of the **fluorescent label**).

5) The **complementary base sequence** can be **read** from the gel. The **smallest** nucleotide (e.g. one base) is at the **bottom** of the gel. Each band after this represents **one more base** added. So by reading the bands **from the bottom** of the gel **to the top**, you can build up the **DNA sequence** one base at a time.

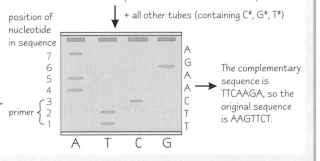

The complementary sequence is TTCAAGA, so the original sequence is AAGTTCT.

Nowadays sequencing is done **altogether** in **one tube** in an **automated DNA sequencer**. The tube contains **all** the modified nucleotides, each with a **different coloured** fluorescent label, and a machine reads the sequence for you.

Practice Questions

Q1   What is a DNA probe?

Q2   Name the two techniques used together to determine the sequence of a gene.

Exam Question

Q1   A piece of DNA 9 kb long is labelled at one end with a fluorescent nucleotide marker. The DNA is then digested using the restriction enzyme *Sal*I. The resulting DNA fragments are separated by electrophoresis to obtain the gel on the right.

     a) How many times did the recognition sequence for *Sal*I appear in the original piece of DNA?   [1 mark]

     b) Use the gel to produce a restriction map of the DNA piece.   [3 marks]

     c) Explain why there are more DNA fragments in the partial digest lane than in the total digest lane.   [2 marks]

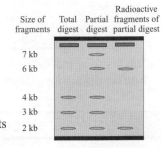

Restriction mapping — I'm not very good with coordinates...

Okay, I won't deny it — there are a couple of difficult bits on this page, like restriction mapping. But just keep drawing the diagrams until the DNA pieces fall into place. All the techniques on this double page are perfect for exam questions, so make sure you learn them well. That way you'll be prepared for anything... in the exam that is.

DNA Probes in Medical Diagnosis

So a scientist could sequence a gene just for the fun of it — or they could do it to help diagnose a genetic disorder.

Many **Human Diseases** are **Caused** by **Mutated Genes**

Some **mutated genes** can cause **diseases** such as **genetic disorders** and **cancer** (see page 85).
Other mutations can produce genes that are **useful** in **some situations** but **not** in others. For example:

Sickle-cell Anaemia

- Sickle-cell anaemia is a **recessive genetic disorder** caused by a **mutation** in the **haemoglobin gene**.
- The mutation causes an **altered haemoglobin protein** to be produced, which makes **red blood cells sickle-shaped** (**concave**).
- The **sickle** red blood cells **block capillaries** and **restrict blood flow**, causing **organ damage** and periods of **acute pain**.
- Some people are **carriers** of the disease — they have **one normal** and **one mutated copy** of the haemoglobin gene.
- Sickle-cell carriers are partially **protected** from **malaria**.
- This **advantageous** effect has caused an increase in the frequency of the sickle-cell **allele** (the mutated version of the gene) in areas where malaria is **common** (e.g. parts of Africa).
- However, this also **increases** the **likelihood** of people in those areas inheriting **two copies** of the sickle-cell allele, which means more people will **suffer** from the disease in these areas.

DNA Probes Can be Used to Screen for Mutated Genes

DNA probes (see page 100) can be used to look (**screen**) for clinically important genes
(e.g. **mutated genes** that result in **genetic disorders**). There are two ways to do this:

1) The probe can be **labelled** and used to look for a **single gene** in a sample of DNA, as shown on page 100.
2) Or the probe can be used as part of a **DNA microarray**, which can screen **lots** of **genes** at the **same time**:

- A **DNA microarray** is a **glass slide** with **microscopic spots** of **different** DNA probes **attached** to it in **rows**.
- A sample of **labelled human DNA** is washed over the array.
- If the labelled human DNA **contains** any **DNA sequences** that **match** any of the **probes**, it will **stick** to the array.
- The array is **washed**, to remove any labelled DNA that **hasn't** stuck to it.
- The array is then **visualised** under **UV light** — any **labelled DNA attached** to a probe will **show up** (fluoresce).
- Any spot that fluoresces means that the person's DNA **contains** that specific **gene**. E.g. if the probe is for a mutated gene that causes a **genetic disorder**, this person has the gene and so **has** the disorder.

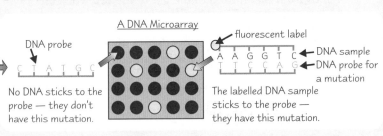

A DNA Microarray

DNA probe
C T A T G C
No DNA sticks to the probe — they don't have this mutation.

fluorescent label
A A G G T C ← DNA sample
T T C C A G ← DNA probe for a mutation
The labelled DNA sample sticks to the probe — they have this mutation.

You also need to know how to **produce** a DNA probe — first the **gene** that you want to screen for is **sequenced** (see previous page). Then **PCR** (see p. 93) is used to produce **multiple copies** of **part** of the gene — these are the **probes**.

Scientific Methods are Continuously Updated and Automated

1) In the **past**, some of the gene technologies you've learnt about on the past few pages were **labour-intensive**, **expensive** and could only be done on a **small scale**.
2) Now these techniques are often **automated**, more **cost-effective** and can be done on a **large scale**.
3) For example, using a single **probe** to **screen** for a single **mutated gene** (see page 100) is **slow** and **small-scale**. Now we have **DNA microarrays** — they're **quick** and can screen for **thousands** of genes at once (see above).
4) Scientific methods like this are **constantly** being **updated** and **automated** to be **faster, cheaper**, and **more accurate** (because there's less human error). This means **medical diagnoses** become **faster** and **more accurate**.

DNA Probes in Medical Diagnosis

The **Results** of **Screening** can be used for **Genetic Counselling**...

1) **Genetic counselling** is **advising patients** and their **relatives** about the **risks of genetic disorders**.

2) It involves **advising** people about **screening** (e.g. looking for mutated genes if there's a **history of cancer**) and **explaining the results** of a screening. Screening can help to **identify the carrier** of a gene, the **type of mutated gene** they're carrying (indicating the type of genetic disorder or cancer) and the **most effective treatment**.

3) If the results of a screening are **positive** (an individual **has** the mutation) then genetic counselling is used to advise the patient on the **options** of **prevention** or **treatment** available. Here are two examples:

   <u>EXAMPLE</u>: A **woman** with a family history of **breast cancer** may have **genetic counselling** to help her **decide** whether or not to be **screened** for **known mutations** that can lead to breast cancer, e.g. a mutation in the BRCA1 **tumour suppressor gene** (see p. 85). If she is screened and the result is **positive**, genetic counsellors might explain that a woman with the mutated BRCA1 gene has an **85%** chance of developing **breast cancer** before the age of **70**. Counselling can also help the woman to **decide** if she wants to take **preventative** measures, e.g. a **mastectomy**, to prevent breast cancer developing.

   <u>EXAMPLE</u>: A couple who are **both carriers** of the **sickle-cell allele** (see previous page) may **like** to have **kids**. They may undergo genetic counselling to help them **understand** the **chances** of them having a child with sickle-cell anaemia (**one in four**). Genetic counselling also provides **unbiased advice** on the possibility of having **IVF** and **screening** their **embryos** for the alleles, so embryos **without the mutation** are **implanted** in the womb. It could also provide information on the **help** and **drugs** available if they have a child with sickle-cell anaemia.

...and **Deciding Treatment**

Cancers can be caused by **mutations** in **proto-oncogenes** (forming **oncogenes**) and mutations in **tumour suppressor genes** (see page 85). **Different mutations** cause **different cancers**, which **respond** to **treatment** in **different ways**. **Screening** using DNA probes for **specific** mutations can be used to help decide the **best** course of **treatment**. For example:

There's more on how identifying the mutation can affect treatment on pages 86-87.

Breast cancer can be caused by a **mutation** in the **HER2 proto-oncogene**. If a patient with breast cancer is screened and tests **positive** for the HER2 oncogene, they can be treated with the drug **Herceptin®**. This drug **binds** specifically to the **altered HER2 protein receptor** and **suppresses cell division**. Herceptin® is **only effective** against this type of breast cancer because it **only targets** the altered HER2 receptor. Studies have shown that **targeted treatment** like this, alongside less-specific treatment like chemotherapy, can **increase survival rate** and **decrease relapse rate** from breast cancer.

Practice Questions

Q1 Give an example of a mutation that is useful in one way but not in another.

Q2 Give an example of a scientific technique that has been automated.

Q3 What is genetic counselling?

Q4 Describe one situation where genetic counselling may be needed.

Exam Questions

Q1 a) Briefly describe how a DNA probe for a clinically important gene can be produced. [2 marks]
   b) Describe how you could screen a person for this gene and many others at the same time. [4 marks]

Q2 Annette has colon cancer. A drug called Cetuximab is used to treat colon cancer caused by a mutation in the KRAS proto-oncogene. Annette is screened and tests negative for the KRAS oncogene.
   a) Why is it unlikely that Annette will be treated with Cetuximab? [1 mark]
   b) Suggest why Annette will undergo genetic counselling. [2 marks]

Information probes — they're called exams...

All of the techniques you've learnt earlier in this section (PCR, sequencing, DNA probes) come together nicely in this medical diagnosis stuff — it's good to know that what you've learnt has a point to it. It's also good to know that as techniques improve, better ways to diagnose some diseases are found.

Gene Therapy

Congratulations — you've made it to the last two pages of the section. I guess I'd better make them good 'uns then...

Gene Therapy Could be Used to Treat or Cure Genetic Disorders and Cancer

How it works:

1) Gene therapy involves **altering** the **defective genes** (mutated alleles) inside cells to treat **genetic disorders** and **cancer**.

2) How you do this depends on whether the disorder is caused by a mutated **dominant allele** or two mutated **recessive alleles** (see page 42):
   - If it's caused by two mutated **recessive** alleles you can **add** a working **dominant allele** to make up for them (you '**supplement**' the faulty ones).
   - If it's caused by a mutated **dominant** allele you can '**silence**' the **dominant allele** (e.g. by sticking a bit of DNA in the middle of the allele so it doesn't work any more).

Gene therapy isn't being used to treat people yet, but some gene therapy treatments are undergoing clinical trials.

A DNA-filled doughnut — surely the best way to deliver new alleles...

How you get the 'new' allele (DNA) inside the cell:

1) The allele is **inserted into cells** using **vectors** (see page 94).

2) Different **vectors** can be used, e.g. altered **viruses**, **plasmids** or **liposomes** (spheres made of lipid).

There are two types of gene therapy:

1) **Somatic therapy** — this involves **altering** the **alleles** in **body cells**, particularly the cells that are **most affected** by the disorder. For example, **cystic fibrosis** (CF) is a genetic disorder that's very **damaging** to the **respiratory system**, so somatic therapy for CF **targets** the epithelial cells lining the lungs. Somatic therapy doesn't affect the individual's **sex cells** (sperm or eggs) though, so any **offspring** could still **inherit** the disease.

2) **Germ line therapy** — this involves **altering** the **alleles** in the **sex cells**. This means that **every cell** of **any offspring** produced from these cells will be **affected** by the gene therapy and they **won't suffer from the disease**. Germ line therapy in humans is currently **illegal** though.

There are Advantages and Disadvantages to Gene Therapy

You need to be able to **evaluate** the **effectiveness** of **gene therapy** — this means being able to discuss the **advantages** and **disadvantages** of the technique, some of which are given in the table below:

ADVANTAGES	DISADVANTAGES
It could prolong the lives of people with genetic disorders and cancer.	The effects of the treatment may be short-lived (only in somatic therapy).
It could give people with genetic disorders and cancer a better quality of life.	The patient might have to undergo multiple treatments (only in somatic therapy).
Carriers of genetic disorders might be able to conceive a baby without that disorder or risk of cancer (only in germ line therapy).	It might be difficult to get the allele into specific body cells.
It could decrease the number of people that suffer from genetic disorders and cancer (only in germ line therapy).	The body could identify vectors as foreign bodies and start an immune response against them.
	An allele could be inserted into the wrong place in the DNA, possibly causing more problems, e.g. cancer.
	An inserted allele could get overexpressed, producing too much of the missing protein.
	Disorders caused by multiple genes (e.g. cancer) would be difficult to treat with this technique.

There are also many **ethical issues** associated with gene therapy. For example, some people are worried that the technology could be used in ways **other** than for **medical treatment**, such as for treating the **cosmetic effects** of **aging**. Other people worry that there's the potential to do **more harm** than good by using the technology (e.g. risk of overexpression of genes — see table).

Gene Therapy

You Might have to **Interpret** Some **Data** on the **Effectiveness** of **Gene Therapy**

In the **exam**, you could get a data **question** on the **effectiveness** of **gene therapy**. So, here's one I did earlier:

Background:

X-linked severe combined immunodeficiency disease (X-linked SCID) is an **inherited disorder** affecting the **immune system**. The disorder is caused by a **mutation** in the **IL2RG gene**, located on the **X chromosome**. The IL2RG gene codes for a **protein** that's **essential** for the **development** of some immune system cells, so the sufferer is **vulnerable** to **infectious diseases** and many **die** in **infancy**.

The study:

A study was designed to investigate the **effectiveness** of **gene therapy** in patients with **X-linked SCID**. **Ten patients** were treated with a **virus vector** carrying a **correct version** of the **IL2RG gene**. After gene transfer, the **patient's immune system** was **monitored** for **at least three years** and noted as **functional** (good) or not. Their **health** was also **monitored** for the same time. Bar charts of the results are shown on the right.

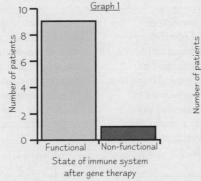

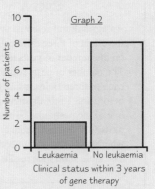

You could be asked to describe the data...

- The **first graph** shows that **nine** out of the **ten** patients had a **functional** immune system **after** gene therapy.

- The **second graph** shows that **two** out of the **ten** patients developed **leukaemia** within 3 years of the treatment.

...or draw conclusions...

- **Gene therapy** can be **used** to **correct** the symptoms of **X-linked SCID**, i.e. produce a **functioning immune system**. However, you **can't** say gene therapy can **cure** X-linked SCID as the study **doesn't** say **how long** the effects of the treatment lasted for.

- **Two out of the ten** patients developed **leukaemia** after the treatment, so there's a chance it's **linked** to the gene therapy (e.g. the vector could have **inserted** the gene into a **proto-oncogene** or **tumour suppressor**, see page 85). But you can only **suggest** a link as **other factors** may have been involved. For example, the patients could have been **more genetically predisposed** to develop cancer.

...or evaluate the methodology

- The **sample size** is **small** — only **ten** patients were treated. This makes the results **less reliable**.

Practice Questions

Q1 How could gene therapy be used to supplement mutated recessive alleles?

Q2 How are supplementary alleles added to human DNA?

Q3 What does germ line gene therapy involve?

Exam Question

Q1 A patient suffering from cystic fibrosis was offered gene therapy targeted at his lung epithelial cells to help treat the disease.

    a) What does gene therapy involve?   [1 mark]

    b) What type of gene therapy was the patient offered?   [1 mark]

    c) Give three possible disadvantages of the treatment.   [3 marks]

First counselling, now therapy — our genes are well looked after...

Now, you might think you need some therapy after this section, but don't worry, the buzzing in your head is normal — it's due to information overload. So go get yourself a cuppa and a biccie and have a break. Then go over some of the difficult bits in this section again. Believe me, the more times you go over it the more things will click into place.

How to Interpret Experiment and Study Data

Science is all about getting good evidence to test your theories... so scientists need to be able to spot a badly designed experiment or study a mile off, and be able to interpret the results of an experiment or study properly. Being the cheeky little monkeys they are, your exam board will want to make sure you can do it too. Here's a quick reference section to show you how to go about interpreting data-style questions.

Here Are Some **Things** You Might be **Asked** to do...

This stuff might be familiar from AS, but you need to know it for A2 as well.

Here are three examples of the kind of data you could expect to get:

Study A

An agricultural scientist investigated the effect of three different pesticides on the number of pests in wheat fields. The number of pests was estimated in each of three fields, using ground traps, before and 1 month after application of one of the pesticides. The number of pests was also estimated in a control field where no pesticide had been applied. The table shows the results.

	Number of pests	
Pesticide	Before application	1 month after application
1	89	98
2	53	11
3	172	94
Control	70	77

Study B

Study B investigated the link between the number of bees in an area and the temperature of the area. The number of bees was estimated at ten 1-acre sites. The temperature was also recorded at each site. The results are shown in the scattergram below.

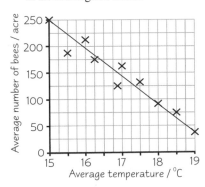

Experiment C

An experiment was conducted to investigate the effect of temperature on the rate of photosynthesis. The rate of photosynthesis in Canadian pondweed was measured at four different temperatures by measuring the volume of oxygen produced. All other variables were kept constant. The results are shown in the graph below.

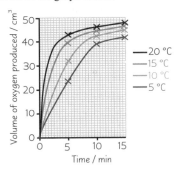

1) Describe and Manipulate the Data

You need to be able to **describe** any data you're given. The level of **detail** in your answer should be appropriate for the **number of marks** given. Loads of marks = more detail, few marks = less detail. You could also be asked to **manipulate** the data you're given (i.e. do some **calculations** on it). For the examples above:

Example — Study A

1) You could be asked to **calculate** the **percentage change** (**increase** or **decrease**) in the number of pests for each of the pesticides and the control. E.g. for pesticide 1: $(98 - 89) \div 89 = 0.10 = $ **10% increase.**

2) You can then use these values to **describe** what the **data** shows — the **percentage increase** in pests in the field treated with **pesticide 1 was the same as for the control** (10% increase) (1 mark). **Pesticide 3 reduced** pest numbers by **45%**, but **pesticide 2 reduced the pest numbers the most** (79% decrease) (1 mark).

Example — Study B

The data shows a **negative correlation** between the average number of bees and the temperature (1 mark).

Correlation describes the **relationship** between two variables — e.g. the one that's been changed and the one that's been measured. Data can show **three** types of correlation:

Positive   Negative   None

1) **Positive** — as one variable **increases** the other **increases**.
2) **Negative** — as one variable **increases** the other **decreases**.
3) **None** — there is **no relationship** between the two variables.

Example — Experiment C

You could be asked to calculate the initial rate of photosynthesis at each temperature: The **gradient = the rate of photosynthesis:**

$$\text{Gradient} = \frac{\text{Change in Y}}{\text{Change in X}}$$

To tell if some data in a table **is correlated** — draw a **scatter diagram** of one variable against the other and **draw a line of best fit.**

How to Interpret Experiment and Study Data

2) Draw or Check a Conclusion

1) Ideally, only **two** quantities would ever change in any experiment or study — everything else would be **constant**.

2) If you can keep everything else constant and the results show a correlation then you **can** conclude that the change in one variable **does cause** the change in the other. ⟹

3) But usually all the variables **can't** be controlled, so other **factors** (that you **couldn't** keep constant) could be having an **effect**.

4) Because of this, scientists have to be very careful when **drawing conclusions**. Most results show a **link** (correlation) between the variables, but that **doesn't prove that a change in one causes the change in the other**. ⟹

Example — Experiment C

All other variables were **kept constant**. E.g. light intensity and CO_2 concentration **stayed the same** each time, so these **couldn't** have influenced the rate of reaction. So you **can say** that an increase in temperature up to 20 °C **causes** an increase in the rate of photosynthesis.

Example — Study B

There's a **negative correlation** between the average number of bees and temperature. But you **can't** conclude that the increase in temperature **causes** the decrease in bees. **Other factors** may have been involved, e.g. there may be **less food** in some areas, there may be **more bee predators** in some areas, or **something else** you hadn't thought of could have caused the pattern...

Example — Experiment C

A science magazine **concluded** from this data that the optimum temperature for photosynthesis is **20 °C**. The data **doesn't** support this. The rate **could** be greatest at 22 °C, or 18 °C, but you can't tell from the data because it doesn't go **higher** than 20 °C and **increases** of **5 °C** at a time were used. The rates of photosynthesis at in-between temperatures **weren't** measured.

5) The **data** should always **support** the conclusion. This may sound obvious but it's easy to **jump** to conclusions. Conclusions have to be **precise** — not make sweeping generalisations. ⟸

3) Explain the Evidence

You could also be asked to **explain** the **evidence** (the data and results) — basically use your **knowledge** of the subject to explain **why** those results were obtained. ⟶

Example — Experiment C

Temperature increases the rate of photosynthesis because it **increases** the **activity** of **enzymes** involved in photosynthesis, so reactions are catalysed more quickly.

4) Comment on the Reliability of the Results

Reliable means the results can be **consistently reproduced** when an experiment or study is repeated. And if the results are reproducible they're more likely to be **true**. If the data isn't reliable for whatever reason you **can't draw** a valid **conclusion**. Here are some of the things that affect the reliability of data:

1) **Size of the data set** — For experiments, the **more repeats** you do, the **more reliable** the data. If you get the **same result** twice, it could be the correct answer. But if you get the same result **20 times**, it's much more reliable. The general rule for **studies** is the larger the **sample size**, the more **reliable** the **data** is.

E.g. Study B is quite **small** — they only studied ten 1-acre sites. The **trend** shown by the data may not appear if you studied **50 or 100 sites**, or studied them for a longer period of time.

2) **The range of values in a data set** — The **closer** all the values are to the **mean**, the **more reliable** the data set.

E.g. Study A is **repeated three more times** for pesticides 2 and 3. The percentage decrease each time is: 79%, 85%, 98% and 65% for **pesticide 2** (**mean = 82%**) and 45%, 45%, 54% and 43% for **pesticide 3** (**mean = 47%**). The data values are **closer to the mean** for **pesticide 3** than pesticide 2, so that data set is **more reliable**. The **spread** of **values about the mean** can be shown by calculating the **standard deviation** (SD). ⟶

The **smaller the SD** the **closer** the values to the **mean** and the **more reliable the data**. SDs can be shown on a graph using **error bars**. The ends of the bars show one SD **above** and one SD **below** the **mean**.

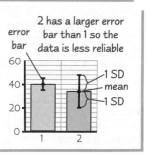

2 has a larger error bar than 1 so the data is less reliable

How to Interpret Experiment and Study Data

3) <u>Variables</u> — The **more variables** you **control**, the **more reliable** your data is. In an experiment you would control all the variables. In a study you try to control **as many as possible**.

The hat, trousers, shirt and tie variables had been well controlled in this study.

E.g. ideally, all the sites in Study B would have a similar **type** of land, similar **weather**, have the same **plants** growing, etc. Then you could be more sure that the one factor being **investigated** (temperature) is having an **effect** on the thing being **measured** (number of bees).

4) <u>Data collection</u> — think about all the **problems** with the **method** and see if **bias** has slipped in.

E.g. in Study A, the traps were placed on the **ground**, so pests like moths or aphids weren't included. This could have affected the results.

5) <u>Controls</u> — without controls, it's very difficult to **draw valid conclusions**. **Negative controls** are used to make sure that nothing you're doing in the experiment has an effect, **other than** what you're testing.

E.g. in Experiment C, the **negative control** would be all the equipment set up as normal but **without** the pondweed. If **no oxygen** was produced at any temperature it would show that the variation in the amount of oxygen produced when there was pondweed was due to the **effect** of temperature on the pondweed, and **not** the effect of temperature on **anything else** in the experiment.

6) <u>Repetition by other scientists</u> — for theories to become accepted as 'fact' other scientists need to **repeat** the work (see page 2). If **multiple studies** or **experiments** come to the same conclusion, then that conclusion is **more reliable**.

E.g. if a second group of scientists repeated Study B and got the same results, the results would be **more reliable**.

There Are a Few *Technical Terms* You *Need to Understand*

I'm sure you probably know these all off by heart, but it's easy to get mixed up sometimes. So here's a quick recap of some words **commonly used** when assessing and analysing experiments and studies:

1) **Variable** — A variable is a **quantity** that has the **potential to change**, e.g. weight. There are two types of variable commonly referred to in experiments:

- **Independent variable** — the thing that's **changed** in an experiment.
- **Dependent variable** — the thing that you **measure** in an experiment.

When drawing graphs, the dependent variable should go on the **y-axis** (the vertical axis) and the independent on the **x-axis** (the horizontal axis).

2) **Accurate** — Accurate results are those that are **really close** to the **true** answer. The true answer is **without error**, so if you can reduce error as much as possible you'll get a more accurate result. The most **accurate methods** are those that produce as **error-free** results as possible.

3) **Precise results** — These are results taken using **sensitive instruments** that measure in **small increments**, e.g. pH measured with a meter (pH 7.692) will be **more precise** than pH measured with paper (pH 8).

It's possible for results to be precise **but not** accurate, e.g. a balance that weighs to 1/1000 th of a gram will give precise results, but if it's not **calibrated** properly the results won't be accurate.

4) **Qualitative** — A **qualitative** test tells you **what's** present, e.g. an acid or an alkali.

5) **Quantitative** — A **quantitative** test tells you **how much** is present, e.g. an acid that's pH 2.46.

There's enough evidence here to conclude that data interpretation is boring...

*These pages should give you a fair idea of how to interpret data. Just use your head and remember the four things you might be asked to do — **d**escribe the **d**ata, **c**heck the **c**onclusions, **e**xplain the **e**vidence and check the **r**esults are **r**eliable.*

Answers

Unit 4: Section 1 — Populations
Page 5 — Populations and Ecosystems

1 a) Maximum of 2 marks available.
They live on farmland, in open woodland, in hedgerows and urban areas **[1 mark]**, and roost in cracks in trees and buildings **[1 mark]**.
  b) Maximum of 3 marks available.
Their wings are light and flexible, which allows them to catch fast and manoeuvrable insects. This increases their chances of catching enough food to survive **[1 mark]**. They use echolocation so they can catch insects that come out at night. This also increases their chances of catching enough food to survive **[1 mark]**. They make unique mating calls so they only attract a mate of the same species. This increases their chance of reproduction by making successful mating more likely **[1 mark]**.
This question is only asking about the biotic conditions (the living features of the ecosystem), so you won't get any marks for talking about abiotic conditions (the non-living features of the ecosystem).

Page 7 — Investigating Populations

1 a) Maximum of 1 mark available.
By taking random samples of the population **[1 mark]**.
  b) Maximum of 4 marks available.
She could use quadrats to measure percentage cover of daffodils **[1 mark]**. Several would be placed on the ground at random locations within the field **[1 mark]**. The percentage of each quadrat that's covered by daffodils would be recorded **[1 mark]**. The percentage cover for the whole field could then be estimated by averaging the data collected in all of the quadrats **[1 mark]**.
Make sure you don't write about quadrats in a transect — they're used to investigate the distribution of plant species, not the percentage cover in an area.

Page 9 — Investigating Populations

1 a) Maximum of 5 marks available.
A group of snails would have been caught **[1 mark]**, marked in a way that wouldn't harm them, e.g. by painting a spot on their shell **[1 mark]**, then released back into the environment **[1 mark]**. After waiting a week a second sample would have been taken **[1 mark]** and the number marked in the second sample would have been counted **[1 mark]**.
  b) Maximum of 2 marks available.

Total population size = $\dfrac{52 \times 38}{14}$ **[1 mark]**

Total population size = 141 **[1 mark]**.
Award 2 marks for correct answer of 141 without any working.

Page 11 — Variation in Population Size

1 a) Maximum of 7 marks available.
In the first three years, the population of prey increases from 5000 to 30 000. The population of predators increases slightly later (in the first five years), from 4000 to 11 000 **[1 mark]**. This is because there's more food available for the predators **[1 mark]**. The prey population then falls after year three to 3000 just before year 10 **[1 mark]**, because lots are being eaten by the large population of predators **[1 mark]**. Shortly after the prey population falls, the predator population also falls (back to 4000 by just after year 10) **[1 mark]**, because there's less food available **[1 mark]**. The same pattern is repeated in years 10-20 **[1 mark]**.
  b) Maximum of 4 marks available.
The population of prey increased to around 40 000 by year 26 **[1 mark]**. This is because there were fewer predators, so fewer prey were eaten **[1 mark]**. The population then decreased after year 26 to 25 000 by year 30 **[1 mark]**. This could be because of intraspecific competition **[1 mark]**.

Page 13 — Human Populations

1 Maximum of 4 marks available.
At stage 1 of the DTM the population size is low and not increasing **[1 mark]**. At stage 5 the population is high but shrinking **[1 mark]**. At stage 1 the population structure is made up of a lot of young people and very few older people **[1 mark]**, but at stage 5 there are few young people and a lot of older people **[1 mark]**.

Unit 4: Section 2 — Energy Supply
Page 15 — Photosynthesis, Respiration and ATP

1 Maximum of 6 marks available, from any of the 8 points below.
In the cell, ATP is synthesised from ADP and inorganic phosphate/P_i **[1 mark]** using energy from an energy-releasing reaction, e.g. respiration **[1 mark]**. The energy is stored as chemical energy in the phosphate bond **[1 mark]**. ATP synthase catalyses this reaction **[1 mark]**. ATP then diffuses to the part of the cell that needs energy **[1 mark]**. Here, it's broken down back into ADP and inorganic phosphate/P_i **[1 mark]**, which is catalysed by ATPase **[1 mark]**. Chemical energy is released from the phosphate bond and used by the cell **[1 mark]**.
Make sure you don't get the two enzymes confused — ATP **syn**thase **syn**thesises ATP, and ATPase breaks it down.

Page 19 — Photosynthesis

1 a) Maximum of 1 mark available.
The thylakoid membranes **[1 mark]**.
  b) Maximum of 1 mark available.
Photosystem II **[1 mark]**.
  c) Maximum of 4 marks available.
Light energy splits water **[1 mark]**.

H_2O **[1 mark]** $\rightarrow 2H^+ + \frac{1}{2} O_2$ **[1 mark]**.

The electrons from the water replace the electrons lost from chlorophyll **[1 mark]**.
The question asks you to explain the purpose of photolysis, so make sure you include why the water is split up — to replace the electrons lost from chlorophyll.
  d) Maximum of 1 mark available.
NADP **[1 mark]**.

2 a) Maximum of 6 marks available.
Ribulose bisphosphate/RuBP and carbon dioxide/CO_2 join together to form an unstable 6-carbon compound **[1 mark]**. This reaction is catalysed by the enzyme rubisco/ribulose bisphosphate carboxylase **[1 mark]**. The compound breaks down into two molecules of a 3-carbon compound called glycerate 3-phosphate/GP **[1 mark]**. Two molecules of glycerate 3-phosphate are then converted into two molecules of triose phosphate/TP **[1 mark]**. The energy for this reaction comes from ATP **[1 mark]** and the H^+ ions come from reduced NADP **[1 mark]**.
  b) Maximum of 2 marks available.
Ribulose bisphosphate is regenerated from triose phosphate/TP molecules **[1 mark]**. ATP provides the energy to do this **[1 mark]**.
This question is only worth two marks so only the main facts are needed, without the detail of the number of molecules.
  c) Maximum of 3 marks available.
No glycerate 3-phosphate/GP would be produced **[1 mark]**, so no triose phosphate/TP would be produced **[1 mark]**. This means there would be no glucose produced **[1 mark]**.

Page 21 — Limiting Factors in Photosynthesis

1 a) Maximum of 4 marks available.
By burning propane to increase air CO_2 concentration **[1 mark]**. By adding heaters to increase temperature **[1 mark]**. By adding coolers to decrease temperature **[1 mark]**. By adding lamps to provide light at night **[1 mark]**.
  b) Maximum of 2 marks available.
Potatoes **[1 mark]** because the yield showed the smallest percentage increase of 25% (850 – 680 = 170, 170 ÷ 680 × 100 = 25%) **[1 mark]**.

Answers

Page 23 — Respiration

1   Maximum of 6 marks available, from any of the 7 points below.
First, the 6-carbon glucose molecule is phosphorylated **[1 mark]**
by adding two phosphates from two molecules of ATP **[1 mark]**.
This creates two molecules of triose phosphate **[1 mark]** and two
molecules of ADP **[1 mark]**. Triose phosphate is oxidised
(by removing hydrogen) to give two molecules of 3-carbon
pyruvate **[1 mark]**. The hydrogen is accepted by two molecules
of NAD, producing two molecules of reduced NAD **[1 mark]**.
During oxidation four molecules of ATP are produced **[1 mark]**.
When describing glycolysis make sure you get the number of molecules
correct — one glucose molecule produces two molecules of triose
phosphate. You could draw a diagram in the exam to show the reactions.

2 a) Maximum of 2 marks available.
Pyruvate + reduced NAD **[1 mark]** → lactate + NAD **[1 mark]**
  b) Maximum of 2 marks available.
Lactate is produced to regenerate NAD **[1 mark]** so glycolysis can
continue and ATP can be produced under anaerobic conditions to
provide energy for biological processes **[1 mark]**.

Page 25 — Respiration

1 a) Maximum of 2 mark available.
The transfer of electrons down the electron transport
chain stops **[1 mark]**. So there's no energy released to
phosphorylate ADP/produce ATP **[1 mark]**.
  b) Maximum of 2 marks available.
The Krebs cycle stops **[1 mark]** because there's no oxidised NAD/FAD
coming from the electron transport chain **[1 mark]**.
Remember that when the electron transport chain is inhibited, the
reactions that depend on the products of the chain are also affected.

Unit 4: Section 3 — Energy Flow and Nutrient Cycles

Page 27 — Energy Transfer and Productivity

1 a) Maximum of 2 marks available.
Gross productivity = net productivity + respiratory loss
$1245 + 4165 = 5410$ **[1 mark]**
gross productivity = $5410 \ kJm^{-2} \ yr^{-1}$ **[1 mark]**
Award 2 marks for correct answer of $5410 \ kJm^{-2} \ yr^{-1}$ without
any working.
  b) Maximum of 3 marks available.
Because not all of the energy is taken in by the Arctic hares **[1 mark]**.
Some parts of the grass aren't eaten, so the energy isn't taken in
[1 mark] and some parts of the grass are indigestible, so they'll pass
through the hares and come out as waste **[1 mark]**.
  c) Maximum of 2 marks available.
$(11 \div 137) \times 100 = 8$ **[1 mark]**
energy transfer efficiency = 8% **[1 mark]**
Award 2 marks for correct answer of 8% without any working.

Page 29 — Farming Practices and Productivity

1   Maximum of 5 marks available.
Organic farmers might use biological agents **[1 mark]**. Biological
agents reduce the numbers of pests, so crops lose less energy and
biomass, increasing productivity **[1 mark]**. They include natural
predators that eat the pest species to reduce their numbers **[1 mark]**.
They could also use parasites that live in or lay their eggs on pest
insects, killing the pests or reducing their ability to function **[1 mark]**.
They could also introduce pathogenic bacteria or viruses that
kill pests **[1 mark]**.

Page 31 — The Carbon Cycle and Global Warming

1   Maximum of 6 marks available.
Carbon from CO_2 in the air and water becomes carbon compounds
in plants when they photosynthesise **[1 mark]**. Carbon is then passed
onto primary consumers when they eat the plants, and secondary and
tertiary consumers when they eat the other consumers **[1 mark]**.
When organisms die, the carbon in the dead organisms is digested
by microorganisms called decomposers **[1 mark]**. Carbon is returned

to the atmosphere as CO_2 because all living organisms carry out
respiration, which produces CO_2 **[1 mark]**. When dead organic
matter ends up in a place where there aren't any decomposers the
carbon can be turned into fossil fuels over millions of years **[1 mark]**.
Carbon in fossil fuels is released back into the atmosphere when
they are burnt **[1 mark]**.

2 a) Maximum of 6 marks available.
Atmospheric CO_2 concentration has increased rapidly since the mid-
19th century due to increased burning of fossil fuels, e.g. for industry
or in cars **[1 mark]**, and the increased destruction of natural CO_2 sinks,
e.g. by deforestation **[1 mark]**. Atmospheric methane concentration
has also been increasing rapidly since the mid-19th century due to
things like increased extraction of fossil fuels **[1 mark]**, more decaying
waste in landfill sites **[1 mark]** and more cattle giving off methane as a
waste gas **[1 mark]**. Methane can also be released when natural stores,
such as frozen ground, thaw **[1 mark]**.
  b) Maximum of 4 marks available.
Global warming is the term for the increase in average global
temperature over the last century **[1 mark]**. Global warming is caused
by enhancement of the greenhouse effect. The greenhouse effect is
when atmospheric greenhouse gases absorb outgoing energy, so less is
lost to space **[1 mark]**. The greenhouse effect helps to keep the planet
warm, but too much greenhouse gas causes the planet to warm up
[1 mark]. Increasing atmospheric concentrations of the greenhouse
gases CO_2 and methane are enhancing the greenhouse effect, and so
causing global warming **[1 mark]**.

Page 33 — Effects of Global Warming

1 a) Maximum of 4 marks available.
CO_2 concentration shows a general trend of increasing **[1 mark]**
from around 338 ppm in 1980 to around 368 ppm in 2000 **[1 mark]**.
The corn yield fluctuates but shows a general trend of
increasing **[1 mark]** from around 105 bushels per acre in
1980 to around 135 bushels per acre in 2000 **[1 mark]**.
  b) Maximum of 1 mark available.
There's a positive correlation between CO_2 concentration and corn
yield / as CO_2 concentration increases so does corn yield **[1 mark]**.
Even though there's a correlation you can't conclude that increasing CO_2
concentration is causing increases in corn yield, because there could be
other factors involved, e.g. changing rainfall pattern.
  c) Maximum of 2 marks available.
CO_2 concentration is a limiting factor for photosynthesis **[1 mark]**,
so increasing CO_2 concentration could mean crops grow faster,
increasing crop yields **[1 mark]**.

Page 35 — The Nitrogen Cycle and Eutrophication

1 a) Maximum of 2 marks available.
A — ammonification **[1 mark]**, C — denitrification **[1 mark]**
  b) Maximum of 3 marks available.
Process B is nitrogen fixation **[1 mark]**. Nitrogen fixation is where
nitrogen gas in the atmosphere is turned into ammonia **[1 mark]**
by bacteria **[1 mark]**.

Unit 4: Section 4 — Succession and Conservation

Page 37 — Succession

1 a) Maximum of 6 marks available.
This is an example of secondary succession, because there is already a
soil layer present in the field **[1 mark]**. The first species to grow will be
the pioneer species, which in this case will be larger plants **[1 mark]**.
These will then be replaced with shrubs and smaller trees **[1 mark]**.
At each stage, different plants and animals that are better adapted for
the improved conditions will move in, out-compete the species already
there, and become the dominant species **[1 mark]**. As succession goes
on, the ecosystem becomes more complex, so species diversity (the
number and abundance of different species) increases **[1 mark]**.
Eventually large trees will grow, forming the climax community,
which is the final seral stage **[1 mark]**.
  b) Maximum of 2 marks available.
Ploughing destroys any plants that were growing **[1 mark]**,
so larger plants may start to grow, but they won't have long enough
to establish themselves before the field is ploughed again **[1 mark]**.

Answers

Page 39 — Conservation

1 a) *Maximum of 6 marks available.*
Conserving rainforests is important for humans as they may provide lots of things that humans need such as clothes, food or drugs [1 mark]. If they're cut down, the source of these things will be lost and they won't be available in the future [1 mark]. Some people think rainforests should be conserved because it's the right thing to do — they think forests have a right to exist, and people don't have a right to cut them all down [1 mark]. The rainforests bring joy to lots of people who visit them. If they're cut down future generations won't be able to enjoy them [1 mark]. Conservation of rainforests will mean less trees are burnt. This helps to prevent climate change because burning trees releases CO_2 into the atmosphere, which contributes to global warming [1 mark]. Conserving rainforests will help to prevent the disruption of food chains. A decrease in one species could mean the loss of many more species in the food chain as there's less food, so more resources are lost [1 mark].

b) *Maximum of 2 marks available, from any of the 3 points below.*
E.g. seedbanks [1 mark], captive breeding programmes [1 mark], relocation [1 mark].

Page 41 — Conservation Evidence and Data

1 a) *Maximum of 4 marks available.*
The cod stock size increased from around 150 000 tonnes in 1963 to around 250 000 tonnes in 1971 [1 mark]. The stock size then decreased (apart from a couple of smaller increases) to around 30 000 tonnes in 2006 [1 mark]. The fishing mortality rate fluctuated but showed a trend of increasing [1 mark] from around 0.5 in 1963 to just below 0.8 in 2006 [1 mark].

b) *Maximum of 2 marks available.*
There's a link between fishing mortality rate and the cod stock size [1 mark]. As the fishing mortality rate increases, the cod stock size decreases/there's a negative correlation between fishing mortality rate and cod stock size [1 mark].

c) *Maximum of 1 mark available.*
1978/1979 [1 mark]

d) *Maximum of 1 mark available.*
It could be used by governments to make decisions about cod fishing quotas (the amount of cod allowed to be removed from the sea by fishermen each year), to try to keep the cod stock above 150 000 tonnes [1 mark].

Unit 4: Section 5 — Inheritance, Selection and Speciation
Page 43 — Inheritance

1 a) *Maximum of 3 marks available.*
Parents' genotypes identified as RR and rr [1 mark]. Correct genetic diagram drawn with gametes' alleles identified as R, R and r, r [1 mark] and gametes crossed to show Rr as the only possible genotype [1 mark].
The question specifically asks you to draw a genetic diagram so make sure that you include one in your answer, e.g.

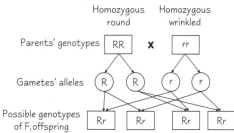

b) *Maximum of 3 marks available.*
Gametes' alleles (produced by F_1 generation) identified as R, r and R, r [1 mark]. Gametes crossed to show genotypes RR, Rr and rr in a 1:2:1 ratio [1 mark]. RR and Rr genotypes identified as giving a round phenotype and rr as wrinkled phenotype, giving a 3:1 ratio of round : wrinkled seeds [1 mark]. Award three marks for a correct ratio of 3:1 for round : wrinkled seeds.

The question doesn't ask for a genetic diagram but it can help you work out the answer, e.g.

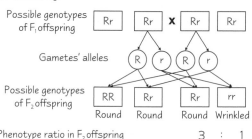

Page 45 — Inheritance

1 *Maximum of 3 marks available.*
Men only have one copy of the X chromosome (XY) but women have two (XX) [1 mark]. Haemophilia A is caused by a recessive allele so females would need two copies of the allele for them to have haemophilia A [1 mark]. As males only have one X chromosome they only need one recessive allele to have haemophilia A, which makes them more likely to have haemophilia A than females [1 mark].

2 *Maximum of 4 marks available.*
Genotypes of parents identified as $I^A I^O$ and $I^B I^B$ [1 mark]. Correct genetic diagram drawn with gametes' alleles identified as I^A, I^O and I^B, I^B [1 mark] and gametes crossed to show genotypes $I^A I^B$ and $I^B I^O$ in a 1:1 ratio [1 mark]. The probability of the couple having a child with blood group B is 0.5 (or 50%) [1 mark].
The question specifically asks you to draw a genetic diagram so make sure that you include one in your answer, e.g.

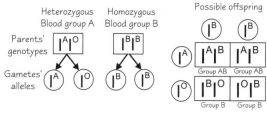

3 a) *Maximum of 1 mark available.*
Individual 2 could have genotype AA or Aa [1 mark].
Individual 2 is an unaffected male so he must have at least one A allele (AA or Aa). But you can't say for sure if he has AA or Aa.

b) *Maximum of 2 marks available*
Individual 6's genotype is Aa [1 mark]. The offspring of individuals 5 and 6 has ADA deficiency (aa), so both parents must be carriers of the recessive allele (Aa) [1 mark]. Or, individual 3 has ADA deficiency (aa), so must have passed a recessive allele onto individual 6. Individual 6 is unaffected so must have a dominant allele as well (Aa) [1 mark].

c) *Maximum of 4 marks available*
Parents' genotypes identified as both Aa [1 mark]. Correct genetic diagram drawn with gametes' alleles identified as A, a and A, a [1 mark] and gametes crossed to show genotypes AA, Aa and aa in a 1:2:1 ratio [1 mark]. aa genotype identified as causing ADA deficiency, giving a probability of 0.25 (25%) [1 mark].
The question asks you to show your working so you could draw a genetic diagram, e.g.

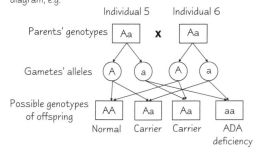

Answers

Page 47 — The Hardy-Weinberg Principle

1 a) Maximum of 3 marks available.
Frequency of genotype CC = p^2 = 0.14 **[1 mark]** so the frequency of the dominant allele C = p = $\sqrt{0.14}$ = 0.37 **[1 mark]**.
The frequency of the recessive allele c = q = 1 − p = 1 − 0.37 = 0.63 **[1 mark]**. Award three marks for a correct answer of 0.63.

b) Maximum of 1 mark available.
Frequency of homozygous recessive genotype cc = q^2 = 0.63^2 = 0.40 **[1 mark]**.

c) Maximum of 2 marks available.
Those that don't have a cleft chin are homozygous recessive cc = 40% **[1 mark]**, so the percentage that do have a cleft chin, Cc or CC, is 100% − 40% = 60% **[1 mark]**. Award two marks for a correct answer of 60%.
There are other ways of calculating this answer, e.g. working out the value of 2pq and adding it to p^2. It doesn't matter which way you do it as long as you get the right answer.

Page 49 — Allele Frequency and Speciation

1 a) Maximum of 4 marks available.
As temperature decreases from 22 °C to 16 °C the frequency of h, the long hair allele, increases from 0.11 to 0.23 **[1 mark]**. This could be because the allele for long hair is more beneficial at colder temperatures **[1 mark]**. Hamsters with the h allele will have a greater chance of surviving, reproducing and passing on their genes, including the beneficial h allele **[1 mark]**. So a greater proportion of the next generation will inherit the beneficial allele and the frequency of the h allele will increase **[1 mark]**.

b) Maximum of 1 mark available.
Directional selection **[1 mark]**.

Unit 5: Section 1 — Responding to the Environment

Page 51 — Nervous and Hormonal Communication

1 a) Maximum of 1 mark available.
Nervous communication is more suitable because electrical impulses travel faster than hormones, so the (protective) response is quicker **[1 mark]**.

b) Maximum of 5 marks available.
Receptors detect the stimulus **[1 mark]**, e.g. light receptors (photoreceptors) in the animal's eyes detect the bright light **[1 mark]**. The receptors send impulses along neurones (via the CNS) to the effectors **[1 mark]**. The effectors bring about a response **[1 mark]**, e.g. the circular iris muscles contract to constrict the pupils and protect the eyes **[1 mark]**.

Page 53 — Receptors

1 Maximum of 6 marks available, from any of the 7 points below.
A tap on the arm is a mechanical stimulus that's detected by pressure receptors/mechanoreceptors called Pacinian corpuscles **[1 mark]**. The stimulus deforms the layers of connective tissue (lamellae) **[1 mark]**, which press on the sensory nerve ending **[1 mark]**. This causes deformation of stretch-mediated sodium ion channels in the neurone cell membrane **[1 mark]**. Sodium ion channels open and sodium ions diffuse into the cell **[1 mark]**. This creates a generator potential **[1 mark]**. If the generator potential reaches the threshold it triggers a nerve impulse/action potential **[1 mark]**.

2 Maximum of 5 marks available.
The human eye has high sensitivity because many rods join one neurone **[1 mark]**, so many weak generator potentials combine to reach the threshold and trigger an action potential **[1 mark]**. The human eye has high acuity because cones are close together and one cone joins one neurone **[1 mark]**. When light from two points hits two cones, action potentials from each cone go to the brain **[1 mark]**. So you can distinguish two points that are close together as two separate points **[1 mark]**.

Page 56 — Nervous System — Neurones

1 a) Maximum of 1 mark available.
Stimulus **[1 mark]**.

b) Maximum of 3 marks available.
A stimulus causes sodium ion channels in the neurone cell membrane to open **[1 mark]**. Sodium ions diffuse into the cell **[1 mark]**, so the membrane becomes depolarised **[1 mark]**.

c) Maximum of 2 marks available.
The membrane was in the refractory period **[1 mark]**, so the sodium ion channels were recovering and couldn't be opened **[1 mark]**.

2 Maximum of 5 marks available.
Transmission of action potentials will be slower in neurones with damaged myelin sheaths **[1 mark]**. This is because myelin is an electrical insulator **[1 mark]**, so increases the speed of action potential conduction **[1 mark]**. The action potentials 'jump' between the nodes of Ranvier/between the myelin sheaths **[1 mark]**, where sodium ion channels are concentrated **[1 mark]**.
Don't panic if a question mentions something you haven't learnt about. You might not know anything about multiple sclerosis but that's fine, because you're not supposed to. All you need to know to get full marks here is how myelination affects the speed of action potential conduction.

Page 59 — Nervous System — Synaptic Transmission

1 Maximum of 6 marks available, from any of the 8 points below.
The action potential arriving at the presynaptic membrane stimulates voltage-gated calcium ion channels to open **[1 mark]**, so calcium ions diffuse into the neurone **[1 mark]**. This causes synaptic vesicles, containing acetylcholine, to fuse with the presynaptic membrane **[1 mark]**. The vesicles release acetylcholine into the synaptic cleft **[1 mark]**. The acetylcholine diffuses across the synaptic cleft **[1 mark]** and binds to cholinergic receptors on the postsynaptic membrane **[1 mark]**. This causes sodium ion channels in the postsynaptic membrane to open **[1 mark]** and the influx of sodium ions triggers a new action potential to be generated at the postsynaptic membrane **[1 mark]**.

2 Maximum of 4 marks available.
They might have weaker muscular responses than normal **[1 mark]**. If receptors are destroyed at neuromuscular junctions then there will be fewer receptors for acetylcholine/ACh to bind to **[1 mark]**, so fewer sodium ion channels will open **[1 mark]**, meaning fewer muscle cells can be stimulated **[1 mark]**.

3 Maximum of 3 marks available.
Galantamine would stop acetylcholinesterase/AChE breaking down acetylcholine/ACh, so there would be more ACh in the synaptic cleft **[1 mark]** and it would be there for longer **[1 mark]**. This means more nicotinic cholinergic receptors would be stimulated **[1 mark]**.

Page 61 — Effectors — Muscle Contraction

1 Maximum of 3 marks available.
Muscles are made up of bundles of muscle fibres **[1 mark]**.
Muscle fibres contain long organelles called myofibrils **[1 mark]**.
Myofibrils contain bundles of myofilaments **[1 mark]**.

2 a) Maximum of 3 marks available.
A = sarcomere **[1 mark]**.
B = Z-line **[1 mark]**.
C = H-zone **[1 mark]**.

b) Maximum of 2 marks available.
The A-bands stay the same length during contraction **[1 mark]**.
The I-bands get shorter **[1 mark]**.

c) Maximum of 3 marks available.
Drawing number 3 **[1 mark]** because the M-line connects the middle of the myosin filaments **[1 mark]**. The cross-section would only show myosin filaments, which are the thick filaments **[1 mark]**.
The answer isn't drawing number 1 because all the dots in the cross-section are smaller, so the filaments shown are thin actin filaments — which aren't found at the M-line.

Answers

Page 63 — Effectors — Muscle Contraction

1   Maximum of 3 marks available.
    Muscles need ATP to relax because ATP provides the energy to
    break the actin-myosin cross bridges [**1 mark**]. If the cross bridges
    can't be broken, the myosin heads will remain attached to the actin
    filaments [**1 mark**], so the actin filaments can't slide back to their
    relaxed position [**1 mark**].

2   Maximum of 3 marks available.
    The muscles won't contract [**1 mark**] because calcium ions won't be
    released into the sarcoplasm, so troponin won't be removed from its
    binding site [**1 mark**]. This means no actin-myosin cross bridges can
    be formed [**1 mark**].

Page 65 — Responses in Animals

1 a) Maximum of 5 marks available.
    High blood pressure is detected by pressure receptors in the aorta
    called baroreceptors [**1 mark**]. Impulses are sent along sensory
    neurones to the medulla [**1 mark**]. Impulses are then sent to the SAN
    along a parasympathetic neurone [**1 mark**]. The parasympathetic
    neurone secretes acetylcholine, which binds to receptors on the
    sinoatrial node/SAN [**1 mark**]. This slows the heart rate (reducing
    blood pressure) [**1 mark**].
  b) Maximum of 2 marks available.
    No impulses sent from the medulla would reach the SAN [**1 mark**],
    so the heart rate wouldn't increase or decrease/control of the
    heart rate would be lost [**1 mark**].

Page 67 — Survival and Responses in Plants

1 a) Maximum of 2 marks available.
    The data shows that the plants provided with auxins grew more than
    those not given auxins [**1 mark**]. This is because auxins stimulate plant
    growth (by cell elongation) [**1 mark**].
  b) Maximum of 5 marks available.
    Auxin is redistributed to the shaded side of the plant [**1 mark**].
    In shoots, auxin stimulates cell elongation [**1 mark**] so the shoots bend
    to grow towards the light [**1 mark**]. In roots, high concentrations of
    auxin inhibit cell elongation [**1 mark**] so the roots bend to grow away
    from the light [**1 mark**].
  c) Maximum of 2 marks available.
    The students should repeat the experiment to see if their results
    are reliable [**1 mark**]. They should also make sure all other variable
    conditions are the same for each group [**1 mark**].
  d) Maximum of 1 mark available.
    The results suggest that auxins stimulate plant growth, so auxins could
    be used to increase tomato yield [**1 mark**].

Unit 5: Section 2 — Homeostasis
Page 69 — Homeostasis Basics

1 a) Maximum of 2 marks available.
    Statement A [**1 mark**] because body temperature continues to increase
    from the normal level and isn't returned [**1 mark**].
  b) Maximum of 2 marks available.
    It makes metabolic reactions less efficient [**1 mark**] because the
    enzymes that control metabolic reactions may denature [**1 mark**].

2   Maximum of 2 marks available.
    Multiple negative feedback mechanisms give more control over
    changes in the internal environment than just having one feedback
    mechanism [**1 mark**]. This is because you can actively increase or
    decrease a level so it returns to normal [**1 mark**].

Page 71 — Control of Body Temperature

1   Maximum of 4 marks available, from the 8 points below. 1 mark for
    each method, up to a maximum of 2 marks. 1 mark for each
    explanation, up to a maximum of 2 marks.
    Vasoconstriction of blood vessels [**1 mark**] reduces heat loss because
    less blood flows through the capillaries in the surface layers of the
    dermis [**1 mark**]. Erector pili muscles contract to make hairs stand
    on end [**1 mark**], trapping an insulating layer of air to prevent
    heat loss [**1 mark**]. Muscles contract in spasms to make the

body shiver [**1 mark**], so more heat is produced from increased
respiration [**1 mark**]. Adrenaline and thyroxine are released [**1 mark**],
which increase metabolism so more heat is produced [**1 mark**].

2   Maximum of 2 marks available.
    Thermoreceptors in the skin detect a higher external temperature
    than normal [**1 mark**]. The thermoreceptors send impulses along
    sensory neurones to the hypothalamus [**1 mark**].

3   Maximum of 4 marks available.
    Snakes are ectotherms [**1 mark**]. They can't control their body
    temperature internally and depend on the temperature of their
    external environment [**1 mark**]. In cold climates, snakes will be
    less active [**1 mark**], which makes it harder to catch prey, avoid
    predators, find a mate, etc. [**1 mark**].
    You need to use a bit of common sense to answer this question —
    you know that the activity level of an ectotherm depends on the
    temperature of the surroundings, so in a cold environment it won't
    be very active. And if it can't be very active it'll have trouble surviving.

Page 73 — Control of Blood Glucose Concentration

1   Maximum of 5 marks available, from the 7 points below.
    High blood glucose concentration is detected by cells in the
    pancreas [**1 mark**]. Beta/β cells secrete insulin into the blood [**1 mark**],
    which binds to receptors on the cell membranes of liver and muscle
    cells [**1 mark**]. This increases the permeability of the cell membranes
    to glucose, so the cells take up more glucose [**1 mark**]. Insulin also
    activates glycogenesis [**1 mark**] and increases the rate that cells
    respire glucose [**1 mark**]. This lowers the concentration of glucose
    in the blood [**1 mark**].
    You need to get the spelling of words like glycogenesis right in the exam
    or you'll miss out on marks.

2   Maximum of 3 marks available.
    They have Type II diabetes [**1 mark**]. They produce insulin, but the
    insulin receptors on their cell membranes don't work properly, so the
    cells don't take up enough glucose [**1 mark**]. This means their blood
    glucose concentration remains higher than normal [**1 mark**].

Page 75 — Control of the Menstrual Cycle

1 a) Maximum of 5 marks available.
    Follicle-stimulating hormone/FSH stimulates a follicle to develop
    [**1 mark**]. FSH also stimulates the ovary to release oestrogen [**1 mark**],
    and oestrogen is also released by the follicle [**1 mark**]. Oestrogen
    inhibits the release of FSH from the anterior pituitary [**1 mark**], so no
    more follicles are stimulated to develop [**1 mark**].
  b) Maximum of 3 marks available.
    High oestrogen concentration stimulates the anterior pituitary to
    release luteinising hormone/LH [**1 mark**]. LH stimulates the ovary
    to release more oestrogen [**1 mark**], which further stimulates the
    anterior pituitary in a positive feedback loop [**1 mark**].

2   Maximum of 3 marks available.
    Oestrogen and progesterone inhibit follicle-stimulating hormone/
    FSH release from the anterior pituitary [**1 mark**]. Because there's
    no FSH, the follicle isn't stimulated to develop [**1 mark**], so there
    is no ovulation [**1 mark**].

Unit 5: Section 3 — Genetics
Page 77 — DNA and RNA

1 a) Maximum of 2 marks available.
    The sugar in DNA is a deoxyribose sugar whilst in RNA it is a ribose
    sugar [**1 mark**]. The bases in DNA are adenine, thymine, guanine
    and cytosine. In RNA, thymine is replaced by uracil [**1 mark**].
    You've only been asked to describe the differences in composition so you
    don't need to write about the shape of the molecules.
  b) i) Maximum of 1 mark available.
       tRNA [**1 mark**]
    ii) Maximum of 1 mark available.
       DNA [**1 mark**]
    iii) Maximum of 1 mark available.
       tRNA [**1 mark**]

Answers

Page 79 — Protein Synthesis

1 Maximum of 2 marks available.
   The drug binds to DNA, preventing RNA polymerase from binding,
   so transcription can't take place and no mRNA can be made [1 mark].
   This means there's no mRNA for translation and so protein synthesis
   is inhibited [1 mark].

2 a) Maximum of 2 marks available.
   $10 \times 3 = 30$ nucleotides long [1 mark]. Each amino acid is coded
   for by three nucleotides (a codon), so the mRNA length in nucleotides
   is the number of amino acids multiplied by three [1 mark].
   b) Maximum of 3 marks available.
   Greater [1 mark]. DNA contains introns and exons [1 mark] but the
   introns are removed to form mRNA by splicing, so the DNA sequence
   will have more nucleotides than the mRNA sequence [1 mark].

Page 81 — The Genetic Code and Nucleic Acids

1 a) Maximum of 2 marks available.
   The mRNA sequence is 18 nucleotides long and the protein produced
   from it is 6 amino acids long [1 mark]. $18 \div 6 = 3$, suggesting three
   nucleotides code for a single amino acid [1 mark].
   b) Maximum of 2 marks available.
   The protein produced was leucine-cysteine-glycine. This would only
   be produced if the code is non-overlapping, e.g. UUGUGUGGG =
   UUG-UGU-GGG = leucine-cysteine-glycine [1 mark].
   If the code was overlapping the codons would be UUG-UGU-GUG-UGU,
   which would give the protein leucine-cysteine-valine-cysteine.
   Also the protein produced is only 6 amino acids long, which is correct
   if the code is non-overlapping — the protein would be longer if the
   code overlapped [1 mark].

2 a) Maximum of 2 marks available. Award 2 marks if all four amino acids
   are correct and in the correct order. Award 1 mark if three amino acids
   are correct and in the correct order.
   GUG = valine
   UGU = cysteine
   CGC= arginine
   GCA = alanine
   Correct sequence = valine, cysteine, arginine, alanine.
   b) Maximum of 2 marks available. Award 2 marks if all four codons are
   correct and in the correct order. Award 1 mark if three codons are
   correct and in the correct order.
   arginine = CGC
   alanine = GCA
   leucine = UUG
   phenylalanine = UUU
   Correct sequence = CGC GCA UUG UUU.
   c) Maximum of 3 marks available.
   valine = GUG
   arginine = CGC
   alanine = GCA
   mRNA sequence = GUG CGC GCA.
   DNA sequence = CAC [1 mark] GCG [1 mark] CGT [1 mark].

Page 83 — Regulation of Transcription and Translation

1 a) Maximum of 4 marks available.
   The results of tubes 1 and 2 suggest that oestrogen affects the
   expression of the gene for the Chi protein [1 mark] because
   mRNA and active protein production only occur in the presence
   of oestrogen [1 mark]. When oestrogen is present it binds to the
   oestrogen receptor (transcription factor), forming an oestrogen-
   oestrogen receptor complex [1 mark]. This complex works as an
   activator, helping RNA polymerase to bind to the DNA, activating
   transcription and resulting in protein production in the presence
   of oestrogen [1 mark].
   b) Maximum of 3 marks available.
   The mutant could have a faulty oestrogen receptor [1 mark].
   Oestrogen might not bind to the receptor, or the oestrogen-oestrogen
   receptor complex might not work as an activator [1 mark]. This would
   mean even in the presence of oestrogen transcription wouldn't be
   activated, so no mRNA or protein would be produced [1 mark].
   This is a pretty tricky question — drawing a diagram of how oestrogen
   controls transcription would help you figure out the answer.

c) Maximum of 3 marks available.
   E.g. the results would be no full length mRNA and no protein
   produced [1 mark]. The siRNA and associated proteins would attach
   to the mRNA of the Chi protein and cut it up into smaller portions,
   resulting in no full length mRNA [1 mark]. No mRNA would be
   available for translation, so no protein would be produced [1 mark].

Page 85 — Mutations, Genetic Disorders and Cancer

1 a) Maximum of 1 mark available.
   AGCTATGAGGCC
   b) Maximum of 5 marks available.
   The original gene codes for the amino acid sequence serine-tyrosine-
   glutamine-alanine [1 mark]. The mutated gene codes for the amino
   acid sequence arginine-tyrosine-glutamic acid-alanine [1 mark].
   Even though there are three mutations there are only two changes to
   the amino acid sequence [1 mark]. This is because of the degenerate
   nature of the DNA code, which means more than one codon can code
   for the same amino acid [1 mark]. So the substitution mutation on the
   last triplet doesn't alter the amino acid (GCT and GCC both code
   for alanine) [1 mark].
   c) Maximum of 2 marks available.
   Acquired [1 mark], because they weren't present before exposure to
   a mutagenic agent / they weren't present in the gametes [1 mark].
   d) Maximum of 3 marks available.
   The mutations could result in the gene becoming an
   oncogene [1 mark]. When they are functioning normally,
   proto-oncogenes stimulate cell division by producing proteins
   that make cells divide [1 mark]. However, when proto-oncogenes
   mutate to form oncogenes the gene can become overactive.
   This stimulates the cells to divide uncontrollably (the rate of
   division increases), resulting in a tumour (cancer) [1 mark].

Page 87 — Diagnosing and Treating Cancer and Genetic Disorders

1 Maximum of 25 marks available.
   HINTS:
   • Start off by describing what hereditary and acquired mutations are and
     the types of disorders they cause, e.g. genetic disorders and cancer.
   • Then explain how knowing that some cancers are caused by acquired
     mutations affects prevention, e.g. if you know that mutagenic agents
     cause acquired mutations you can try to avoid them and so help
     prevent cancer.
   • Next do the same for how knowing that some cancers are caused by
     acquired mutations affects diagnosis, e.g. high risk individuals can be
     screened more frequently to try to diagnose cancer earlier. Don't forget
     to include that early diagnosis increases their chances of recovery.
   • Then repeat the previous two bits for hereditary cancer and genetic
     disorders.
   • Use plenty of examples from pages 86-87 to back up your points.

Page 89 — Stem Cells

1 a) Maximum of 1 mark available.
   Totipotent cells are stem cells that can mature into any cell type in
   an organism [1 mark].
   b) Maximum of 3 marks available.
   The totipotent cells grew and divided at all pHs, but they grew the
   most at pH 4 (up to 30 g in mass) [1 mark]. The totipotent cells
   only specialised into shoot cells at pH 4 [1 mark]. This suggests
   that the pH helps to control the specialisation of cells for this type
   of plant [1 mark].

Page 91 — Stem Cells in Medicine

1 Maximum of 4 marks available.
   E.g. stem cell therapies are currently being used for some diseases
   affecting the blood and immune system [1 mark]. Bone marrow
   contains stem cells that can become specialised to form any type of
   blood cell [1 mark]. Bone marrow transplants can be used to replace
   faulty bone marrow in patients with leukaemia (a cancer of the blood
   or bone marrow) [1 mark]. The stem cells in the transplanted bone
   marrow divide and specialise to produce healthy blood cells [1 mark].

Answers

2 Maximum of 2 marks available.
Obtaining embryonic stem cells involves the destruction of an embryo *[1 mark]*. Some people believe that embryos have a right to life and that it's wrong to destroy them *[1 mark]*.

Unit 5: Section 4 — Gene Technology
Page 93 — Making DNA Fragments

1 a) Maximum of 2 marks available.
There's a BamHI recognition sequence at either side of the DNA fragment, so you could use this restriction endonuclease to isolate the fragment *[1 mark]*. BamHI would be incubated with the bacterial DNA, so that it cuts the DNA at each of these recognition sequences *[1 mark]*.
You could include a diagram to help explain your answer.

b) Maximum of 5 marks available, from any of the 6 points below.
The bacterial DNA is mixed with free nucleotides, primers and DNA polymerase *[1 mark]*. The mixture is heated to 95 °C to break the hydrogen bonds *[1 mark]*. The mixture is then cooled to between 50 – 65 °C to allow the primers to bind/anneal to the DNA *[1 mark]*. The primers bind/anneal to the DNA because they have a sequence that's complementary to the sequence at the start of the DNA fragment (e.g. CAA) *[1 mark]*. The mixture is then heated to 72 °C and DNA polymerase lines up free nucleotides along each template strand, producing new strands of DNA *[1 mark]*. The cycle would be repeated over and over to produce lots of copies *[1 mark]*.
This question asks you to describe and explain, so you need to give the reasons why each stage is done to gain full marks.

Page 95 — Gene Cloning

1 a) Maximum of 2 marks available.
Colony A is visible/fluoresces under UV light, but Colony B isn't visible/ doesn't flouresce *[1 mark]*. So only Colony A contains the fluorescent marker gene, which means it contains transformed cells *[1 mark]*.

b) Maximum of 3 marks available.
The plasmid vector DNA would have been cut open with the same restriction endonuclease that was used to isolate the DNA fragment containing the target gene *[1 mark]*. The plasmid DNA and gene (DNA fragment) would have been mixed together with DNA ligase *[1 mark]*. DNA ligase joins the sticky ends of the DNA fragment to the sticky ends of the plasmid DNA *[1 mark]*.

Page 97 — Genetic Engineering

1 a) Maximum of 3 marks available.
The drought-resistant gene could be inserted into a plasmid *[1 mark]*. The plasmid is then inserted into a bacterium *[1 mark]*, which is used as a vector to get the gene into the plant cells *[1 mark]*.

b) Maximum of 2 marks available.
The transformed wheat plants could be grown in drought-prone regions *[1 mark]*, where they would reduce the risk of famine and malnutrition *[1 mark]*.

c) Maximum of 1 mark available.
They could be concerned that the large agricultural company will have control over the recombinant DNA technology used to make the drought-resistant plants, which could force smaller companies out of business *[1 mark]*.

Page 99 — Genetic Fingerprinting

1 a) Maximum of 5 marks available, from any of the 8 points below.
A sample of DNA is obtained, e.g. from a person's blood/saliva/ skin etc. *[1 mark]*. PCR is used to amplify multiple areas containing different sequence repeats *[1 mark]*, using primers that anneal to either side of a repeat so the whole repeat is amplified *[1 mark]*. A fluorescent tag is added to all the DNA fragments *[1 mark]*. The DNA mixture undergoes electrophoresis *[1 mark]* — the DNA mixture is placed into a well in a slab of gel and an electric current is applied *[1 mark]*. The DNA fragments move towards the positive electrode at the other end of the gel and separate out *[1 mark]*. The separated bands produce the genetic fingerprint, which is viewed under UV light *[1 mark]*.

b) Maximum of 2 marks available.
Genetic fingerprint 1 is most likely to be from the child's

father *[1 mark]* because five out of six of the bands on his genetic fingerprint match that of the child's, compared to only one on fingerprint 2 *[1 mark]*.

c) Maximum of 1 mark available, from any of the 4 points below.
E.g. they can be used to link a person to a crime scene (forensic science) *[1 mark]*. To prevent inbreeding between animals or plants *[1 mark]*. To diagnose cancer or genetic disorders *[1 mark]*. To investigate the genetic variability of a population *[1 mark]*.

Page 101 — Locating and Sequencing Genes

1 a) Maximum of 1 mark available.
Two *[1 mark]*.
Three fragments were produced in the total digest, so it must have cut one piece of DNA in two places.

b) Maximum of 3 marks available. 1 mark for 1 fragment in the correct place, 2 marks for 2/3 fragments in the correct place and 1 mark for labelling the Sal1 restriction sites.

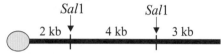

You can tell that the first cut must be after 2 kb because it's the smallest radioactive fragment in the partial digest. You can tell that the next fragment is 4 kb and not 3 kb because the other radioactive fragment in the partial digest is 6 kb long (2 kb + 4 kb). If the middle fragment was 3 kb then the radioactive fragment in the partial digest would be 5 kb long (2 + 3).

c) Maximum of 2 marks available.
Because Sal1 has not been left long enough to cut at all of its recognition sequences *[1 mark]*, so there are other lengths of DNA present, i.e. 2 + 4 = 6 kb, 4 + 3 = 7 kb *[1 mark]*.

Page 103 — DNA Probes in Medical Diagnosis

1 a) Maximum of 2 marks available.
The gene that you want to screen for is sequenced *[1 mark]*. Multiple copies of parts of the gene are made by PCR to be used as DNA probes *[1 mark]*.

b) Maximum of 4 marks available.
Microscopic spots of different DNA probes are attached in series to a glass slide, producing a microarray *[1 mark]*. A sample of the person's labelled DNA is washed over the array and if any of the DNA matches any of the probes, it will stick to the array *[1 mark]*. The array is washed and visualised, under UV light/X-ray film *[1 mark]*. Any spot that shows up means that the person's DNA contains that specific gene *[1 mark]*.
This question asks you to describe how many genes can be screened for at once (which is a microarray), but you could be asked how you can use DNA probes to look for a single gene too (see page 100).

2 a) Maximum of 1 mark available.
Because she tested negative for the mutated gene (KRAS oncogene) that the drug specifically targets *[1 mark]*.

b) Maximum of 2 marks available.
So the results of her screening can be explained to her *[1 mark]* and so her treatment options can also be explained *[1 mark]*.

Page 105 — Gene Therapy

1 a) Maximum of 1 mark available.
Gene therapy involves altering/supplementing defective genes (mutated alleles) inside cells to treat genetic disorders and cancer *[1 mark]*.

b) Maximum of 1 mark available.
Somatic gene therapy *[1 mark]*.

c) Maximum of 3 marks available, from any 6 of the points below.
E.g. the effect of the treatment may be short-lived *[1 mark]*. The patient might have to undergo multiple treatments *[1 mark]*. It might be difficult to get the allele into specific body cells *[1 mark]*. The body may start an immune response against the vector *[1 mark]*. The allele may be inserted into the wrong place in the DNA, which could cause more problems *[1 mark]*. The allele may become overexpressed *[1 mark]*.

Index

Index

Index